ANIMALS AND BUSINESS

This book explores the role, impact and experiences of animals within organisations and the wider socioeconomic contexts in which they participate.

Animals are vital to business. From agriculture to tourism, logistics to digital media, animals work, reproduce, create value and support livelihoods. Yet business and management theory still largely treats organisations as purely human domains. *Animals and Business* challenges this view. By introducing an interspecies approach to responsible management, the book explores how animals are not just resources or external ethical concerns, but key actors within organisations and industries. It examines how business practices shape animals' lives and how, in turn, animals influence how value is created and sustained. Each chapter takes an animal-centred approach to a familiar idea such as work, consumption and innovation. Case studies range from broiler chickens in global supply chains to bees in forest honey economies and medical scent-detection dogs in Nordic healthcare. Developed in dialogue with international scholars, the book's central claim is simple: Business is better understood, and better practised, when we recognise the animals whose lives and labour are intertwined with it.

This book is essential reading for students and researchers working at the intersection of business studies, animal studies and sustainability.

Helen Wadham is a Reader in Sustainability and Head of Doctoral College in the Faculty of Business and Law at Manchester Metropolitan University, UK.

José-Carlos García-Rosell is a Professor of Business Ethics in Oulu Business School at the University of Oulu, Finland.

John Lever is a Reader in Sustainable and Resilient Communities in the Faculty of Business and Law at Manchester Metropolitan University, UK.

ANIMALS AND BUSINESS

An Interspecies Approach to Responsible Management

Helen Wadham, José-Carlos García-Rosell and John Lever

Routledge
Taylor & Francis Group
LONDON AND NEW YORK

earthscan
from Routledge

Designed cover image: Onfokus

First published 2027
by Routledge
4 Park Square, Milton Park, Abingdon, Oxon OX14 4RN

and by Routledge
605 Third Avenue, New York, NY 10158

Routledge is an imprint of the Taylor & Francis Group, an informa business

British Library Cataloguing-in-Publication Data
A catalogue record for this book is available from the British Library

ISBN: 978-1-032-95570-4 (hbk)
ISBN: 978-1-032-95565-0 (pbk)
ISBN: 978-1-003-58554-1 (ebk)

DOI: 10.4324/9781003585541

Typeset in Optima
by Taylor & Francis Books

CONTENTS

ABOUT THE AUTHORS

Helen Wadham is a Reader in Sustainability at Manchester Metropolitan University, UK. Her research examines responsible management and the purpose of business. She has a particular interest in human–animal relations, interspecies work and care, informal organising and community-led approaches to change. She is a Fellow of the Royal Anthropological Institute.

José-Carlos García-Rosell works as a Professor of Business Ethics at the Oulu Business School, University of Oulu, Finland. He works in the domains of corporate social responsibility, business ethics, and responsible management education. His current research agenda is centred on four interconnected areas: responsible tourism business, human–animal relations in organisations, non-human stakeholders, and experiential learning.

John Lever is a Reader in Sustainable and Resilient Communities at Manchester Metropolitan University, UK. His research focuses on religious food markets and sustainable food systems, a key theme across these strands of work being farm animal welfare and interspecies relations. John is a sociologist by training and a fellow of the Norbert Elias Foundation.

EXPERT COLLABORATORS

Nona Borgström is a PhD student at the Faculty of Veterinary Medicine, Department of Equine and Small Animal Medicine at the University of Helsinki, Finland. She is especially interested in multispecies and multiprofessional interactions in social and healthcare settings, scent detection training, One Welfare and ethics.

Kate Dashper is a Professor in the Carnegie School of Sport and the Centre of Social Justice in Sport and Society at Leeds Beckett University, UK. Her research combines expertise in gender studies and human–animal studies in the contexts of events, leisure, tourism and sport.

David Fennell is a Professor of Geography and Tourism Studies at Brock University, Canada. He has written widely on ecotourism, tourism ethics and moral issues tied to the use of animals in tourism. He is Editor-in-Chief of the *Journal of Ecotourism*.

Anna Hielm-Björkman is an Adjunct Professor and Senior Clinical Teacher at the Department of Equine and Small Animal Medicine, University of Helsinki, Finland. Her research centres on chronic pain, adjunct therapies, dog health and nutrition, and scent detection dog topics.

Lewis Holloway is a Professor in the School of Environmental and Life Sciences at the University of Hull, UK. His research has focused on the production of different knowledges associated with farming and food; on the implications of technological interventions in agriculture; on human–nonhuman relationships in rural spaces; and on "alternative" ways of living in the countryside and of producing and distributing food. He is the lead editor of the *Elgar Encyclopedia of Food and Society*.

Astrid Huopalainen is an Assistant Professor (tenure-track) in the field of Leadership for Creativity, in the Department of Management Studies and the Department of Art and Media at Aalto University. Her research is in Animal Organisation Studies, organisation theory, and critical management studies, with a particular interest in arts-based, posthumanist, practice-based, and processual approaches to issues of human–animal work and everyday organising.

Helen Kopnina is an Assistant Professor at Newcastle Business School, University of Northumbria, UK. She is the author of over 200 peer-reviewed articles and (co)author and (co)editor of 17 books on interrelated topics of environmental sustainability, sustainable business, circular economy, biological conservation, environmental ethics, animal ethics and environmental education. She holds a PhD from the University of Cambridge.

Janet Lowore is Director of Programmes at Bees for Development, UK. She leads international initiatives using beekeeping to support livelihoods, reduce deforestation and advance the Sustainable Development Goals, and holds a PhD in forest beekeeping.

Carlos Monterrubio is a social scientist based at the Universidad Autónoma del Estado de México, Mexico. His research focuses on the sociocultural dimensions of gender, sexuality and multispecies relationships in tourism, leisure and events in Latin America, particularly in Mexico. He holds a PhD from the Manchester Metropolitan University, UK.

Emma Roe is a Professor in More Than Human Geographies and a Fellow of the Academy of Social Sciences. She is a world-leading trans-disciplinary more-than-human geographies scholar within the field of animal geographies and agri-food studies. Recent research projects on food safety, antimicrobial resistance and food system resilience have each involved working alongside food industry businesses and policymakers.

Salini Saha is an independent scholar with a PhD in Sociology from Presidency University, Kolkata, India. She was formerly a Senior Research Associate at National Institute of Advanced Studies, Bengaluru, India, conducting interdisciplinary research on gifted education. Her broader research work spans multispecies ethnography, cultural geography and human–animal relations in eastern India, foregrounding relational and decolonial approaches.

Doris Schneeberger has been a Research Associate at the Institute for Change Management and Management Development at the Vienna University of Economics and Business, Austria since 2019. Her research focuses on the role of nonhuman animals in organisations, integrating topics from organisational studies, animal ethics, psychoanalysis and the ethics of artificial intelligence.

Meghan Shaw is a Research Fellow at BehaviourWorks Australia, Monash University. She is focused on understanding why we do the things that we do. Her interests lie in communication, conservation and biodiversity, and designing messages to motivate people to act for the environment.

Linda Tallberg works as a Senior Lecturer at the University of Lapland, Finland. Her research has exclusively focused on human–animal relations in business and society on questions of ethics, care and justice. Her work revolves around creating a post-anthropocentric agenda in business, management and organisation that includes nonhuman interests.

Dinesh Wadiwel is an Associate Professor in human rights and socio-legal studies at the University of Sydney, Australia. He has a background in social and political theory and is an expert in human–animal studies and disability rights. He has a background working within civil society organisations, including in anti-poverty and disability rights roles.

ACKNOWLEDGEMENTS

We're very grateful to all our expert collaborators – Nona Borgström, Kate Dashper, David Fennell, Anna Hielm-Björkman, Lewis Holloway, Astrid Huopalainen, Helen Kopnina, Janet Lowore, Carlos Monterrubio, Emma Roe, Salini Saha, Doris Schneeberger, Meghan Shaw, Linda Tallberg and Dinesh Wadiwel. It has been a real pleasure to work alongside you, and this book is much richer because of your insight, creativity and generosity. We hope it reflects the spirit and substance of your contributions.

INTRODUCTION

Animals in business and society: Present yet overlooked

Animals are deeply embedded in the everyday workings of business and organisations. They are bred, transported, trained, confined, monitored, displayed, marketed and – in many cases – put to work. Chickens move through global supply chains, horses take part in sport, leisure and tourism industries, dogs are deployed in policing, security and care work, and fish are managed, harvested and traded across extractive and regenerative economies. Yet despite this extensive involvement in economic activity, animals are typically positioned at the margins of how business and society are conceptualised, governed and taught (Donaldson & Kymlicka, 2011; Tallberg et al., 2022a). Within dominant frameworks in economics, management and public policy, animals are treated primarily as resources, background conditions or externalities, rather than as integral components of social and economic systems whose lives and interests are shaped by – and help shape – economic arrangements (Tallberg & Hamilton, 2022). This chapter explores the human-centred assumptions that underpin conventional understandings of organisations and offers an interspecies alternative that recognises animals as active participants in organisational life.

The marginalisation of animals reflects long-standing assumptions in economics, management and public policy about human exceptionalism (Regan, 1983; Singer, 1975). These assumptions determine who and what counts as a legitimate actor or concern, placing animals outside the usual moral and analytical boundaries of mainstream management scholarship (Tallberg & Hamilton, 2022). Even where these fields address responsibility, sustainability or value, animals are typically treated as things to be managed rather than as beings who organise their own lives and relationships while also being actively enrolled into human systems of organisation (Coulter, 2016).

DOI: 10.4324/9781003585541-1

This human-centric way of thinking is increasingly difficult to maintain. Across contemporary economies, animals are central to how organisations operate and how value is produced and contested. In organisational settings, they are not merely symbolic or passive resources: They work, reproduce, consume and resist within arrangements designed primarily around human goals, often without visibility, consent or care (Coulter, 2016; Hamilton & Taylor, 2017). As Despret (2016) argues, animals do not simply occupy the roles assigned to them but actively shape the situations in which they are involved, responding, negotiating and sometimes disrupting the arrangements that seek to contain them. Attending to animals in business requires asking how they are organised and how they participate in and modify organisational life. To overlook this is to overlook how many sectors actually function and how harms and responsibilities are distributed.

Recent scholarship has therefore argued that animals should be understood not as peripheral to business activity, but as participants, workers and subjects within organisational and economic systems (Coulter, 2016; Tallberg et al., 2022b). This shift is visible in growing work on animals as workers, non-human stakeholders and objects of governance, as well as sites of ethical and political contestation (García-Rosell, 2022; Jammaers & Huopalainen, 2023). Such perspectives challenge familiar distinctions between nature and economy, production and care, and ethics and efficiency (Wadham, 2020), while drawing attention to the unequal power relations through which animals are enrolled into business practices, often without representation or the possibility of refusal (Coulter & Fitzgerald, 2019; Donaldson & Kymlicka, 2011).

Ignoring animals is therefore not only a moral issue, although the ethical stakes are significant. Rather, it also limits our analytical and practical capacity to address some of the most pressing challenges facing business and society (Tallberg & Hamilton, 2022). Industrial animal agriculture, for example, is deeply implicated in climate change, biodiversity loss and public health risks, from intensive livestock systems to biosecure processing plants, yet remains marginal within mainstream business education and corporate strategy (Gerber et al., 2013; Whitton et al., 2021). Similarly, debates around innovation and sustainability often prioritise technical or market-based solutions while overlooking the multispecies relationships through which these problems are produced and experienced (Janssens, 2023; Whelan & Gond, 2017). Attending to animals helps reveal blind spots in dominant theories of work, consumption, governance and value, while bringing questions of accountability and justice into sharper focus by showing how the treatment of animals is often connected to wider patterns of inequality and exclusion (Blattner et al., 2020; Jammaers, 2023).

By taking animals seriously, scholarship in this field can develop a more complete account of contemporary organisational life. Paying attention to the animals who already share our workplaces, supply chains and cities strengthens the field's ability to engage with intersecting social, ecological and economic challenges, and deepens insight into how organisations function, where they struggle, and how they might be reorganised in more liveable and responsible ways (Tallberg et al., 2022b, 2024).

Having established the extent to which animals are embedded yet often marginalised in contemporary business, the remainder of this introduction proceeds in five steps. First, we situate the discussion within the broader "animal turn" and the development of Animal Organisation Studies. Second, we examine how anthropocentrism is reproduced within business education and why this matters pedagogically. Third, we clarify the book's aims and conceptual orientation, including our use of key terms such as interspecies and multispecies. Fourth, we reflect on the collaborative approach through which the book was developed. Finally, we explain the structure of the book and how it can be read and used in teaching and learning contexts.

The academic context: The animal turn and its uneven uptake in business and management

Across the social sciences and humanities, the past two to three decades have seen what is often described as the animal turn. This development is most visible in fields such as history, geography and anthropology, where scholars have challenged human-centred accounts of social life and foregrounded animals as subjects shaped by power, labour and governance (Cudworth, 2001; Deckha, 2012). While this shift has reshaped several disciplines, its uptake within business and management – understood here as research and teaching concerned with organisations, markets and managerial practice, spanning organisation studies and more mainstream disciplines such as strategy, marketing, operations and finance – has been slower and more uneven.

Despite animals' long-standing presence in organisational contexts, explicit attention to animals as organisational subjects has emerged primarily in the past decade (Connolly & Cullen, 2018; Labatut et al., 2016). This relatively late engagement reflects the difficulty of unsettling anthropocentric premises in a field closely tied to managerial, economic and instrumental logics.

Within this body of scholarship, engagement with the animal turn has crystallised most clearly in the small but growing subfield of Animal Organisation Studies, or AOS (Tallberg & Hamilton, 2022). Rather than treating animals as external stakeholders or downstream ethical concerns, AOS brings animals directly into the analysis of organising itself. Organisations are reframed as multispecies arrangements in which animals are routinely present, governed and affected through everyday practices (Labatut et al., 2016). They are thus understood not only as entities impacted by organisational decisions, but as beings already embedded within processes of management, labour and value creation.

Empirical studies examine a range of organisational settings, including industrial farming and slaughterhouses, where animals are managed as productive bodies (Hamilton & McCabe, 2016); tourism businesses that rely on animal labour and knowledge (Dashper & Wadham, 2026); and workplaces such as policing, where animals work alongside humans in highly regulated roles (Knight & Sang, 2020). Research also extends to wild animals, whose lives are shaped through conservation management, wildlife governance and tourism regulation (Paring, 2025; Welden, 2023). Across these contexts, animals appear not merely as passive resources but as workers, commodities, dependents or co-actors whose bodies, behaviour and welfare influence organisational outcomes (Cunha et al., 2019; Lever & Evans, 2017). This scholarship raises questions about how value is created, whose labour counts, and how responsibility and care are distributed across species.

At the same time, scholars note that this work remains marginal within the wider field. Even in critical or sustainability-oriented research, animals are often framed instrumentally – as inputs, risks or branding concerns – rather than as beings with interests of their own (Connolly & Cullen, 2018; Smart, 2022). For instance, studies of "pet-friendly" workplaces frequently focus on employee wellbeing while paying limited attention to how animals experience these arrangements (Cunha et al., 2019). Animals therefore tend to matter insofar as they affect human performance, reputation or profit, rather than as organisational subjects in their own right.

Business is nevertheless a particularly important – and challenging – site for engaging the animal turn. It is important because organisations and markets are central mechanisms through which animals are bred, traded, worked and killed at scale, making business a key arena in which multispecies harms are normalised and reproduced (Hamilton & McCabe, 2016). It is challenging because the field remains shaped by anthropocentric and growth-oriented assumptions that prioritise efficiency, control and profitability, and that sit uneasily with challenges to human dominance or managerial authority (Kopnina et al., 2025). These conditions help explain why animal-focused scholarship often struggles for legitimacy within mainstream journals, curricula and research agendas.

Taken together, this literature shows that while the animal turn has begun to reshape business and management scholarship, its uptake remains partial and contested. AOS has opened important conceptual and empirical space for centring animals within organisational analysis, yet it continues to operate at the margins of the field. This unevenness reflects not only disciplinary habits, but deeper questions about what kinds of lives, relationships and responsibilities business scholarship is willing – or able – to recognise.

Business schools, anthropocentrism and the work of absence

Business schools play a central role in preparing future managers, leaders and decision-makers. Through curricula, cases and pedagogical conventions, they frame how students come to understand organisations, value, responsibility and success (Wickert et al., 2021). Hereafter, we use "business education" to refer broadly to teaching and learning within business and management, including organisation studies and related disciplines, foregrounding business as a social and ethical practice rather than merely a technical or instrumental activity.

Graduates of business schools move into sectors deeply entangled with nonhuman animals, including agriculture, food production, tourism, pharmaceuticals, logistics and land-use governance. Yet despite these entanglements, animals remain largely invisible or marginal within business education (Tallberg et al., 2022a). This absence is significant because business education does not simply transmit knowledge but shapes managerial common sense (Tourish, 2020). Decisions about what is treated as "core" influence how future organisational actors recognise responsibility and harm. When animals are absent from dominant frameworks, the effects of organising on nonhuman lives are rendered peripheral, even though managerial decisions routinely structure their working, living and dying conditions (Connolly & Cullen, 2018; Labatut et al., 2016).

At the heart of this pattern lies a largely unspoken anthropocentric assumption about organisations and organising. Mainstream business education tends to treat organisations as exclusively human social systems populated by managers, employees, consumers and stakeholders whose interests are assumed to matter in comparable ways (Calás & Smircich, 2023). Animals, if they appear at all, are framed as resources, commodities, inputs or external ethical concerns rather than as organisational subjects (Pedersen, 2025). This assumption rarely requires explicit defence: It operates as a background condition shaping what is considered relevant, teachable and legitimate within business education, often reinforcing human-centred notions of value and progress (Kopnina, 2018).

Importantly, this marginalisation is not just an oversight. Treating animals as organisational subjects unsettles familiar distinctions between labour and life, production and care, and means and ends (Hamilton & Taylor, 2017; Wadham, 2020). It raises questions about exploitation, dependency and responsibility that sit uneasily with dominant managerial logics of efficiency, growth and control. As work in AOS has shown, organising has always been multispecies, even when animals are rendered invisible or treated instrumentally (Imbrogiano, 2024; Labatut et al., 2016). The persistence of anthropocentrism in business education therefore reflects a stabilising disciplinary move that protects established ways of knowing organisations.

These assumptions become material through curricula and teaching practices. Textbooks, syllabi, cases and teaching materials shape what students come to recognise as "normal" business practice. Across core subjects such as strategy, operations, finance and marketing, animals are routinely absent. Where they do appear, they are often folded into existing frameworks – such as sustainability, consumption or ethics – in ways that leave underlying assumptions intact (Pedersen, 2025). Animals may be discussed as commodities, risks or ethical dilemmas, but rarely as beings whose lives are actively organised through business practices.

Absence, in this sense, does active work. By not appearing in cases and examples, animals are implicitly marked as irrelevant to managerial reasoning (Nocella et al., 2014). Students learn, often without noticing, that nonhuman lives do not belong to the conceptual "core" of business thinking. Responsibility becomes framed narrowly around human stakeholders, while multispecies harms are displaced elsewhere or treated as external to organisational concerns (Sitka-Sage et al., 2017). Ethical reasoning is channelled towards compliance, reputation or consumer sentiment, rather than towards deeper questions of interdependence, care and justice across species (Tallberg et al., 2022a, 2022b).

Pedagogy therefore emerges as a crucial ethical and strategic intervention point. Teachers are not mere transmitters of knowledge, but active participants in shaping how future organisational actors understand their responsibilities and the worlds they inhabit (Pedersen, 2025; Wadham et al., 2023). Choices about what is included, excluded or backgrounded in the classroom contribute to the futures students learn to imagine as possible or desirable. Challenging anthropocentrism in business education is thus not merely a curricular adjustment but an ethical task, requiring educators to reconsider whose lives are rendered visible and valuable through teaching (Kopnina, 2018; Nocella et al., 2014).

Seen in this light, business education is a form of world-making rather than simple knowledge transmission (Wadham et al., 2023). Incorporating animals into business education is not about adding another topic to an already crowded curriculum, but about questioning the assumptions that organise what counts as business knowledge in the first place (Tallberg et al., 2022a). It is at this pedagogical junction – where anthropocentric assumptions, curricular absences and future organisational practices intersect – that this book intervenes.

What this book does: Aims and orientation

Against the background outlined above, this book seeks to reframe how business and management are studied, taught and practised. It brings animals into view as part of institutional worlds by effectively reworking concepts that we tend to take for granted. In doing so, it builds on recent work on multispecies organising and questions of care in place (Ehrnström-Fuentes et al., 2025), while speaking to a growing body of scholarship rethinking animals' place within organisation studies (Huopalainen, 2022; Kandel et al., 2025). Each chapter begins with a concept familiar to readers – such as consumption, innovation or governance – and asks what changes when these are viewed from the standpoint of animals' lives and experiences. In doing so, the book questions rather than simply extends established business frameworks.

Together, the chapters pursue three connected aims that show how animals alter what we can see, what we can think, and what we can and should take responsibility for within business and management. First, the book enables readers to understand how business practices shape animals' lives, and how animals, in turn, influence organising processes and outcomes. Animals appear not as abstract ethical concerns or external stakeholders, but as beings whose bodies, labour and capacities are enrolled within – and actively influence – everyday business decisions. Drawing on research developed in dialogue with educators, practitioners and experts working in animal-related workplace settings, the chapters examine how they are bred, trained, worked with, governed, cared for and harmed through organisational practices (Hamilton & McCabe, 2016; Knight & Sang, 2020). They also attend to how these beings respond, resist, cooperate and co-produce value within these settings. By situating animals as participants in organisational analysis rather than passive recipients of managerial action, the book challenges the assumption that business activity unfolds in a human-only world.

Second, the book shows how adopting an interspecies perspective unsettles taken-for-granted "human" concepts within business and management. Concepts such as work, labour, value, leadership, care and responsibility are often treated as exclusively human concerns. When animals are recognised as organisational subjects, these categories become open to rethinking (Coulter, 2016; Hamilton & Taylor, 2017). Several chapters demonstrate how attention to animals complicates distinctions between production and care, economic and moral value, and means and ends, revealing the limits of anthropocentric management frameworks.

Third, the book foregrounds the ethical and practical issues raised by working and learning alongside other animals. Organisational relationships with animals are not only technical or economic, but deeply ethical, involving questions of dependency, vulnerability, exploitation and responsibility. The chapters examine how animals are positioned within organisational hierarchies, how harm is normalised or obscured, and how ethical responsibility is distributed – or displaced – across species (Connolly & Cullen, 2018; Kopnina, 2018). In doing so, the book moves beyond abstract debates about animal ethics to consider how ethical questions emerge within concrete business settings.

A sustained attention to teaching and learning runs throughout the book. Although Chapter 13 focuses explicitly on pedagogy, concerns about learning and responsibility are woven across the chapters. Teaching is understood not as neutral transmission, but as a practice that constitutes what comes to matter (Wadham et al., 2023). What students are taught to notice, question or ignore influences how they later understand organisations, responsibility and ethical action. Reflective questions and activities therefore invite readers to consider how animals are made visible or invisible in business practice, and how assumptions about value, work and care are learned and reproduced (Kopnina, 2018). In this way, the book positions business education as a strategic site through which organisational futures are imagined and configured.

The book's argument also rests on several key terms that require clarification, namely multispecies, interspecies and animals. Across the social sciences and humanities, "multispecies" has become a widely used term. Quite literally meaning "many species," it signals that social and organisational life involves more than one species and that analysis should not focus on humans alone (Tsing, 2015). The term is often valued for its openness because it acknowledges the presence of multiple species without specifying in advance how they are related.

While multispecies foregrounds the presence of many species, this book adopts an explicitly *"interspecies"* orientation. The prefix "inter" means "between" or "among," and signals a focus on relations rather than simple coexistence. An interspecies approach therefore directs attention to how humans and other animals are brought into contact, positioned in relation to one another, and shaped through practices of care, control, dependency and violence. Instead of simply noting that many species are present, it asks how those relations are organised, governed and contested within specific contexts (Livingstone & Puar, 2011; Tallbear, 2011). This relational emphasis is particularly important for business and management, where organising is fundamentally about structuring relationships, responsibilities and hierarchies. It allows us to examine how business practices actively produce particular forms of human–animal relations – through markets, metrics, routines and institutional norms – rather than merely acknowledging that multiple species are involved (Celermajer et al., 2023).

This orientation also informs our use of the term animals rather than nonhuman animals. The phrase "nonhuman animals" is widely used to remind readers that humans are also animals and to challenge human exceptionalism. However, the prefix "non-" continues to define other animals in relation to the human, subtly reinforcing a human–animal binary that positions humans as the unmarked norm and other animals as deviations from it. By using the term *animals*, this book seeks to unsettle that human–animal binary and to decentre the human as the unmarked reference point. This does not deny important differences between species, nor does it collapse humans and other animals into sameness. Rather, it reflects a commitment to examining how hierarchies and boundaries are constructed and maintained within collective life (Calás & Smircich, 2023; Deckha, 2012).

The book also draws on posthumanist scholarship, which questions the assumption that humans occupy the natural centre of social and economic life (Deckha, 2012). Rather than positioning humans as the default reference point, posthuman approaches situate them within wider networks of living beings. This does not collapse differences between species, since asymmetries of power, vulnerability and responsibility remain central. Instead, it reframes these differences as relational and organised rather than given.

Taken together, these terminological commitments clarify the analytical stance of the book. They signal a relational approach that questions hierarchy and examines how responsibility is organised across species in business life.

Writing differently: Collaboration, openness and how knowledge is made

Bringing animals into business and organisational analysis challenges not only what we study, but how we do scholarship. Animals do not sit comfortably within dominant business frameworks or academic genres designed around human actors, managerial problems and technical solutions. As a result, this book required careful reflection on how knowledge about animals is produced, shared and taught (Labatut et al., 2016; Tallberg et al., 2022a).

From the outset, the book was conceived as a collaborative project rather than a conventional co-authored monograph or edited collection. Each chapter is written by the authors but developed in conversation with one or more "expert collaborators" whose research anchors the chapter's focus. These collaborators did not contribute stand-alone chapters. Instead, they shaped the work through interviews, extended conversations and feedback as arguments evolved. We sought collaborators across different geographical regions, disciplines and career stages. This was not just a matter of inclusivity, but a recognition that emerging fields are shaped through dialogue, experimentation and the cultivation of a wide range of voices. By working across career stages and institutional locations, the book aims not only to synthesise scholarship but to participate in the field's ongoing development. Academic fields are reproduced through decisions about whose work is amplified and legitimised: This project makes those decisions visible and deliberate.

This approach reflects a core commitment of the book. If animals are enrolled into organising through relationships of care, control, dependence and violence, then knowledge about animals should also be produced relationally rather than extractively (Kociatkiewicz & Kostera, 2024; Tallberg & Huopalainen, 2024). Working with expert collaborators helped us remain attentive to ongoing debates and contested practices in specific sectors, while resisting the temptation to simplify ethical tensions. For readers, this means that some chapters may feel deliberative rather than definitive: That is, arguments are shaped through engagement, and certain tensions remain visible rather than resolved.

The book spans diverse organisational contexts, including agriculture, tourism, conservation, food systems and sustainability governance. This breadth reflects the varied ways animals are organised through business practices, often in fragmented and contradictory ways (Hamilton & McCabe, 2016; Kandel et al., 2025). It also reflects the diversity of scholarly approaches to studying animals in and around organisations.

Our collaborators work from different philosophical, ethical and political positions. Some chapters engage reformist efforts within existing systems, while others adopt more critical or oppositional stances. We did not seek to impose a single ethical or theoretical voice across the book. Doing so would have flattened important differences in how animals are understood, valued and governed in practice, and would have misrepresented both the field and the realities of working with animals in organisational settings.

This plurality is, however, uneven. Some species, sectors and geographical contexts are better represented than others, reflecting broader structural inequalities in research funding, institutional access and academic visibility (Deckha, 2012; Tallberg et al., 2024). Much of the scholarship shaping this field has emerged from institutions in the Global North, and this book is inevitably shaped by those academic locations. At the same time, the volume includes contributions grounded in African, Latin American and Asian contexts, reflecting a deliberate effort to widen the geographical scope of the conversation. We acknowledge these limits throughout, recognising that they cannot be fully resolved within a single project.

Working across empirical sites and ethical positions inevitably generated moments of disagreement and misalignment. A recurring challenge lay in maintaining enough coherence for the book to function as a pedagogical resource, while allowing each chapter to retain its own distinct perspective and ethical stance. Rather than smoothing over these differences, we treat them as analytically and pedagogically productive (Stengers, 2018). The book does not present a single, unified account of what "responsible" relations with animals should look like. Instead, it invites engagement with perspectives that are sometimes complementary and sometimes in tension.

Finally, this commitment to openness shapes how the book is likely to be read and used. We do not offer a settled managerial blueprint. Each chapter includes reflective questions designed for classroom discussion and for conversations beyond it. We offer these as a way of continuing the dialogue the book begins, recognising that understanding animals in business is not a closed academic exercise but part of an ongoing social and organisational conversation. The plurality of voices encountered in the book, and in the discussions it generates, is therefore not something to resolve but something to engage with as part of learning to think, teach and organise differently with other animals.

The book's overarching structure and argument

Animals and Business is organised into five Parts (see Figure 0.1), which together develop the book's cumulative argument. We move from interrogating anthropocentric foundations, through reworking core management concepts in light of interspecies relations, to reflecting on institutional responsibility, pedagogy and organisational transformation. This structure reflects the book's central claim: Animals are not an add-on to business thinking but constitutive of how business is organised, valued and taught. For this reason, concepts such as welfare and religion sit alongside more conventionally business-oriented topics, as these domains materially shape how animals are valued, governed and incorporated into organisational life.

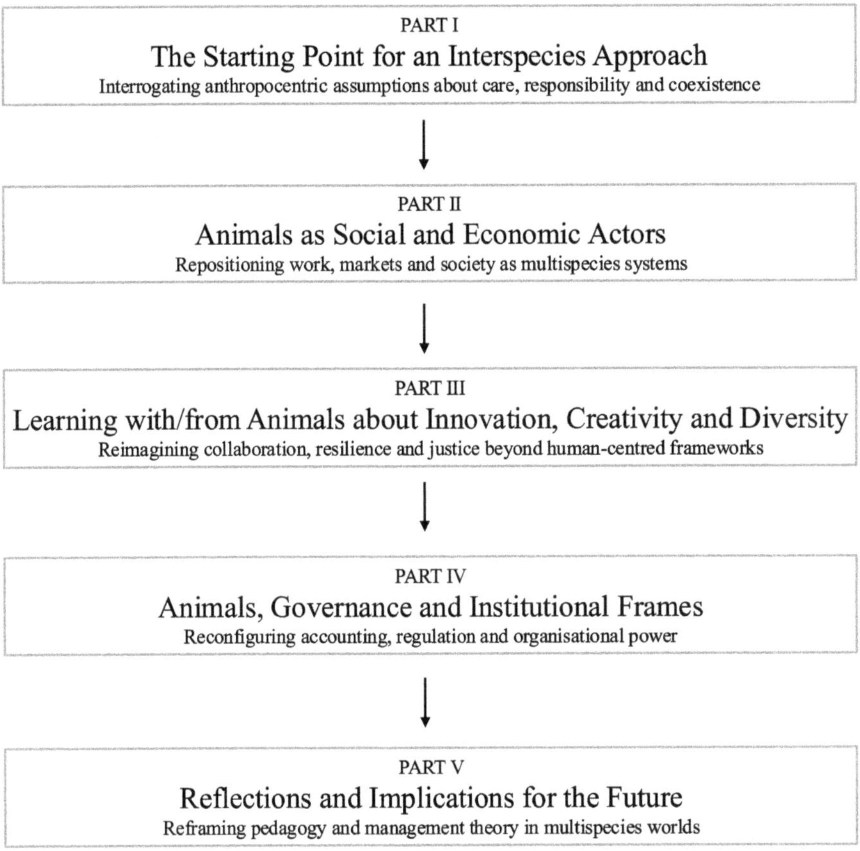

FIGURE 0.1 Conceptual architecture of the book, showing how each part contributes to the reconstitution of management through a multispecies lens.

While the book has a clear conceptual progression, it can also be read in multiple ways. Readers may follow it sequentially, trace a theme across chapters (such as work, value or care), or use individual chapters as stand-alone teaching resources. Each chapter includes reflective questions designed for students, educators and wider audiences, supporting use across undergraduate, postgraduate and executive education contexts.

For lecturers, the book is intended as a flexible pedagogical resource rather than a fixed syllabus. Chapters can be paired with conventional management texts, used to disrupt familiar topics or taught independently within ethics, sustainability or critical management courses. For students and other readers, each chapter offers a starting point for thinking further about the issues raised and how they connect to everyday encounters with animals in business.

Part I: The starting point for an interspecies approach

Part I lays the conceptual groundwork by interrogating human-centred assumptions and showing how responsibility, care and coexistence are negotiated through everyday entanglements with animal life. Chapter 1 explores animals and sustainable development in collaboration with Helen Kopnina. Drawing on the case of elephants in conservation and tourism, it reveals tensions at the heart of sustainability, where protecting nature can involve sacrificing animal welfare. In Chapter 2, Lewis Holloway's research on cattle and sheep farming in northern England examines how illness is managed through systems of measurement and control while still relying on embodied practices of care. The chapter highlights persistent tensions between welfare, productivity and responsibility in contemporary agriculture. Chapter 3 turns to religion through Salini Saha's work on sacred cobras in rural West Bengal, India. Here, animals emerge not merely as symbols within belief systems but as active participants in shaping spiritual practice, moral responsibility and everyday coexistence.

Part II: Animals as social and economic actors

Part II repositions animals as constitutive participants in economic and social systems, challenging the assumption that markets, labour and value are exclusively human domains. Chapter 4 engages animals and consumption through Emma Roe's research on broiler chickens in UK and global meat supply chains. The chapter reframes consumption as a multispecies metabolic process and demonstrates how narratives of "ethical" and "sustainable" eating often depend on rendering animal lives and environmental harm invisible. In Chapter 5, Kate Dashper's work on horses in tourism and leisure settings reconceptualises work as an interspecies activity shaped by cooperation, dependence and unequal power relations. Rather than treating animals as mere tools of service provision, the chapter foregrounds their labour and agency. Chapter 6 turns to digital contexts, drawing on Meghan Shaw's analysis of wildlife campaigns on social media. It shows how animals become visible actors in online spaces, while also revealing how digital visibility can simultaneously support conservation and encourage new forms of exploitation.

Part III: Learning with/from animals about innovation, creativity and diversity

Part III shifts from critique towards possibility, exploring how attention to interspecies relations opens new ways of thinking about collaboration, resilience and justice. Chapter 7 examines animals and innovation through research by Nona Borgström, Anna Hielm-Björkman, Astrid Huopalainen and Linda Tallberg. Focusing on medical scent-detection dogs in Finland, it challenges anthropocentric understandings of innovation and raises ethical questions about welfare, consent and power in collaborative healthcare settings. In Chapter 8, Janet Lowore's work on human–bee relations in forest honey economies in Zambia and Ethiopia reframes creativity as something that emerges through interspecies cooperation rather than human ingenuity alone. The chapter highlights how resilience develops through shared adaptation between people, bees and landscapes. Chapter 9 addresses animals and diversity, equity and inclusion through Carlos Monterrubio's research on working donkeys in rural Mexico. It reveals how exclusion operates across species and argues that justice and inclusion must be rethought as interspecies concerns.

Part IV: Animals, governance and institutional frames

Part IV turns to the institutional level, examining how dominant systems of valuation and regulation structure human–animal relations and how these frameworks might be reimagined. Chapter 10 explores animals and value through Dinesh Wadiwel's analysis of accounting systems in animal agriculture. It shows how animals are treated simultaneously as labour, commodities and raw materials, while their suffering is translated into neutral metrics that obscure ethical concerns. In Chapter 11, David Fennell's work on animal-based tourism governance highlights the persistence of human-only policy frameworks. The chapter argues for recognising animals as legitimate stakeholders within planning and regulatory processes. Chapter 12 concludes this section by examining animals and company activism through Doris Schneeberger's research on Transfarmation Austria and agricultural enterprises transitioning away from animal-based production. It demonstrates how activism can move beyond symbolic gestures towards transforming business models and organisational practices.

Part V: Reflections and implications for the future

Part V concludes the book with two reflective chapters. Chapter 13 considers the pedagogical implications of teaching and learning about business in interspecies worlds. Chapter 14 draws together the book's central insights, reflecting on their significance for business, management and responsibility while acknowledging the partial and ongoing nature of this work.

Drawing these threads together: An invitation to think differently

This book does not offer a definitive account of animals in business and management. It presents a necessarily partial and situated set of perspectives shaped by particular species, sectors, places and scholarly traditions, as well as by the limits of time and access. Some animals, industries and regions appear more prominently than others, and certain tensions remain unresolved. We see this not as a weakness but as a reflection of the field itself: Relations between humans and other animals are complex, contested and uneven, and any attempt at totalising explanation would risk oversimplifying that complexity.

At the same time, the book is designed for use in teaching and learning. Each chapter begins from familiar business concepts – such as work, consumption and innovation – and brings these into conversation with animals' lives and experiences. This structure enables learners to start from what they know while questioning the assumptions that shape that knowledge. The cases, reflective questions and activities are intended not merely as illustrations, but as prompts for discussion, discomfort and ethical reflection. In this sense, the book is about animals, but it is also about widening how responsibility and management are understood.

Across its chapters, the book invites readers to think otherwise about business and management. Thinking with animals unsettles familiar categories: Who is recognised as a worker, what counts as value, where responsibility begins and ends, and how harm is distributed. Animals appear as participants in collective life whose presence reveals forms of dependence, vulnerability and power that stretch across species boundaries. Recognising these dynamics opens space for questioning what appears normal, efficient or inevitable in institutional worlds.

Finally, the book advances an interspecies approach to business and management. It does not treat animals as an add-on to existing theories. Instead, it rethinks core concepts of organisation, responsibility and governance through attention to animals and their relations with humans. In doing so, it brings together insights from Animal Organisation Studies, ethics, geography, anthropology and pedagogy in a form accessible to students and educators. Its central claim is straightforward: Business and management cannot be adequately understood without actively engaging with the animals whose lives are shaped by organisational decisions and who, in turn, shape how organisations function.

Taken together, the chapters do not offer a single answer to how humans should relate to other animals in business contexts. Instead, they provide conceptual and pedagogical tools for seeing differently, teaching differently and asking different questions. In this way, the book aims to support learners and educators in developing more critical, attentive and responsible ways of engaging with the multispecies worlds that business already inhabits. We hope you find it engaging, interesting and useful.

References

Blattner, C. E., Coulter, K., & Kymlicka, W. (2020). *Animal labour: A new frontier of interspecies justice?* Oxford University Press.

Calás, M. B., & Smircich, L. (2023). *A research agenda for organization studies, feminisms and new materialisms*. Edward Elgar.
Celermajer, D., Schlosberg, D., Rickards, L., Stewart-Harawira, M., Thaler, M., Tschakert, P., & Verlie, B. (2023). *Multispecies justice*. Cambridge University Press.
Connolly, L., & Cullen, J. G. (2018). Animals and organisations: An ethic of care framework. *Organization & Environment*, 31(4), 406–424. doi:10.1177/1086026617714329.
Coulter, K. (2016). *Animals, work, and the promise of interspecies solidarity*. Palgrave Macmillan.
Coulter, K., & Fitzgerald, A. (2019). The compounding feminization of poverty and animal exploitation: A multispecies analysis. *Gender, Work & Organization*, 26(3), 288–302. doi:10.1111/gwao.12259.
Cudworth, E. (2001). *Environment and society*. Routledge.
Cunha, M. P. E., Rego, A., & Munro, I. (2019). Dogs in organisations. *Human Relations*, 72(4), 778–800. doi:10.1177/0018726718775524.
Dashper, K., & Wadham, H. (2026). "They treat you like an animal:" Navigating entanglements and productive exclusions within the human/animal boundary in organisations. *Organization*, 33(1), 100–117.
Deckha, M. (2012). Toward a postcolonial, posthumanist feminist theory. *Hypatia*, 27 (3), 527–545. doi:10.1111/j.1527-2001.2012.01291.x.
Despret, V. (2016). *What would animals say if we asked the right questions?* University of Minnesota Press.
Donaldson, S., & Kymlicka, W. (2011). *Zoopolis: A political theory of animal rights*. Oxford University Press.
Ehrnström-Fuentes, M., Böhm, S., Hagolani-Albov, S., & Tesfaye, L. A. (2025). Multispecies organising in the web of life: Ethico-political dynamics of matters of care in ecologies-in-place. *Business Ethics Quarterly*, 35(4), 525–554.
García-Rosell, J. C. (2022). Husky kennels as animal welfare activists: Multispecies relationships as drivers of institutional change. In L. Tallberg & L. Hamilton (Eds), *The Oxford handbook of animal organization studies* (pp. 397–408). Oxford University Press.
Gerber, P. J., Steinfeld, H., Henderson, B., Mottet, A., Opio, C., Dijkman, J., Falcucci, A., & Tempio, G. (2013). *Tackling climate change through livestock: A global assessment of emissions and mitigation opportunities*. FAO.
Hamilton, L., & McCabe, D. (2016). "It's just a job:" Understanding emotion work, animal suffering and resistance. *Organization*, 23(3), 330–349. doi:10.1177/1350508414559987.
Hamilton, L., & Taylor, N. (2017). *Ethnography after humanism*. Palgrave Macmillan.
Huopalainen, A. (2022). Writing with the bitches. *Organization*, 29(6), 959–978. doi:10.1177/13505084221079144.
Imbrogiano, J. P. (2024). De-anthropocentrification now: A call for problematising revisions of human–animal positions in social thought. *Sustainability: Science, Practice and Policy*, 20(1), 2366620. doi:10.1080/15487733.2024.2366620.
Jammaers, E. (2023). On ableism and anthropocentrism: A canine perspective on the workplace inclusion of disabled people. *Human Relations*, 76(2), 233–257. doi:10.1177/00187267221095542.
Jammaers, E., & Huopalainen, A. (2023). "I prefer working with mares, difficult in character but go the extra mile:" A study of multiple inequalities in equine (sports) business." *Gender, Work & Organization*, 30(6), 2049–2068. doi:10.1111/gwao.13044.
Janssens, M. (2023). Animal business, a blind spot of companies. In W. Dubbink & W. van der Deijl (Eds), *Business ethics: A philosophical introduction* (pp. 245–259). Springer.

Kandel, I. J., Dlouhy, K., & Schmitt, A. (2025). Animal roles in organisations: A framework for exploring organisational human–animal relations. *Organization*, 32(4), 508–526.

Knight, C., & Sang, K. (2020). "At home he's a pet, at work he's a colleague." *Culture and Organization*, 26(5–6), 355–371. doi:10.1080/14759551.2019.1609987.

Kociatkiewicz, J., & Kostera, M. (2024). Writing differently: On the constraints and possibilities of presenting research rooted in feminist epistemologies. *Gender, Work & Organization*, 31(1), 284–304. doi:10.1111/gwao.13072.

Kopnina, H. (2018). Education for the future? *Environmental Education Research*, 24 (6), 1–17. doi:10.1080/13504622.2017.1303820.

Kopnina, H., Wong, R., Black, K., & Đurović, M. (2025). Anthropocentrism and climate change: Radical reorientation away from greenwashing and toward degrowth in business education. In S. Brechin & S. Lee (Eds), *Routledge handbook of climate change and society* (pp. 193–207). Routledge.

Labatut, J., Munro, I., & Desmond, J. (2016). Animals and organisations. *Organization*, 23 (3), 315–329.

Lever, J., & Evans, A. (2017). Corporate social responsibility and farm animal welfare: Towards sustainable development in the food industry? In S. Idowu & S. Vertigans (Eds), *Stages of corporate social responsibility*. CSR, Sustainability, Ethics & Governance (pp. 205–222). Springer.

Livingstone, J., & Puar, J. (2011). Interspecies. *Social Text*, 29(1), 3–14. doi:10.1215/01642472-1210217.

Nocella, A., Sorenson, J., Socha, K., & Matsuoka, A. (Eds). (2014). *Defining critical animal studies*. Peter Lang.

Paring, G. (2025). Multispecies ethics and space: Coexisting with wolves. *Journal of Business Ethics*, 200, 255–268. doi:10.1007/s10551-024-05914-6.

Pedersen, H. (2025). Post-anthropocentric pedagogies: Purposes, practices, and insights for higher education. *Teaching in Higher Education*, 30(2), 344–358. doi:10.1080/13562517.2023.2222087.

Regan, T. (1983). *The case for animal rights*. University of California Press.

Singer, P. (1975). *Animal liberation*. HarperCollins.

Sitka-Sage, M. D., Kopnina, H., Blenkinsop, S., & Piersol, L. (2017). Rewilding education in troubled times; or, getting back to the wrong post-nature. *Visions for Sustainability*, 8, 20–37. doi:10.13135/2384-8677/2334.

Smart, J. (2022). Animals as stakeholders. In N. Thomas (Ed.), *Animals and business ethics* (pp. 297–320). Palgrave Macmillan.

Stengers, I. (2018). *Another science is possible*. Polity.

Tallbear, K. (2011). *Native American DNA: Tribal belonging and the false promise of genetic science*. University of Minnesota Press.

Tallberg, L., & Hamilton, L. (2022). *The Oxford handbook of animal organization studies*. Oxford University Press.

Tallberg, L., & Huopalainen, A. (2024). We are the missing people: On posthumanist onto-epistemologies in organisation studies. In F. X. de Vaujany, S. Gherardi, & P. Silva (Eds), *Organisation studies and posthumanism* (pp. 29–52). Routledge.

Tallberg, L., Välikangas, L., & Hamilton, L. (2022a). Animal activism in the business school: Using fierce compassion for teaching critical and positive perspectives. *Management Learning*, 53(1), 55–75. doi:10.1177/13505076211044612.

Tallberg, L., García-Rosell, J. C., & Haanpää, M. (2022b). Human–animal relations in business and society: Advancing the feminist interpretation of stakeholder theory. *Journal of Business Ethics*, 180(1), 1–16. doi:10.1007/s10551-021-04820-2.

Tallberg, L., Huopalainen, A., & García-Rosell, J. C. (2024). Beyond anthropocentrism: A call to action for multispecies inclusivity. *Business & Society*, 64(7), 1263–1268. doi:10.1177/00076503241271254.

Tourish, D. (2020). *Management studies in crisis: Fraud, deception and meaningless research*. Cambridge University Press.

Tsing, A. (2015). *The mushroom at the end of the world: On the possibility of life in capitalist ruins*. Princeton University Press.

Wadham, H. (2020). Horse matters: Re-examining sustainability through human-domestic animal relationships. *Sociologia Ruralis*, 60(3), 530–550. doi:10.1111/soru.12293.

Wadham, H., Hart, C., Hashmi, A., Kettleborough, H. M., Marron, R., Randles, S., … & Tucker, M. (2023). Sustainability education beyond the classroom: How the "exploding university" nurtures collective intelligence across local and global communities. In E. Meletiadou (Ed.), *Handbook of research on implications of sustainable development in higher education* (pp. 202–229). IGI Global.

Welden, E. (2023). Conceptualising multispecies collaboration: Work, animal labour, and nature-based solutions. *Transactions of the Institute of British Geographers*, 48 (3), 541–555. doi:10.1111/tran.12593.

Whelan, G., & Gond, J.-P. (2017). Meat your enemy: Animal rights, alignment, and radical change. *Journal of Management Inquiry*, 26(2), 123–138. doi:10.1177/1056492616649442.

Whitton, C., Bogueva, D., Marinova, D., & Phillips, C. J. C. (2021). Are we approaching peak meat consumption? *Animals*, 11(12), 3466. doi:10.3390/ani11123466.

Wickert, C., Post, C., Doh, J., Prescott, J., & Prencipe, A. (2021). Management research that makes a difference: Broadening the meaning of impact. *Journal of Management Studies*, 58(2), 297–320. doi:10.1111/joms.12666.

PART I

The starting point for an interspecies approach

1

ANIMALS AND SUSTAINABLE DEVELOPMENT

Introduction

Sustainable development is a central pillar of global policy, shaping how governments, organisations and institutions respond to environmental and social challenges (Ruggerio, 2021). This chapter critically examines how animals are positioned – or more often excluded – from global conversations about sustainability. Drawing on the work of expert collaborator Helen Kopnina, we argue instead that sustainable development must expand to include animals as stakeholders, ethical subjects and co-inhabitants of the planet.

In recent years, the United Nations Sustainable Development Goals (SDGs), for example, have become a key reference point for imagining the future of the planet and its inhabitants. The SDGs aim to end poverty, reduce inequality, improve health and protect the environment by 2030. Yet despite their inclusive language, they prioritise human wellbeing, economic growth and technological progress, offering little acknowledgement of the lives or welfare of nonhuman animals. When animals appear, they are framed mainly as resources – livestock or biodiversity assets – instead of sentient beings (Kopnina, 2020). For example, SDG 14 ("Life below water") refers to rebuilding fish stocks but overlooks the sentience and welfare of fish. Similarly, SDG 15 ("Life on land") focuses on halting biodiversity loss but makes no reference to the wellbeing of the terrestrial wildlife it seeks to protect.

This anthropocentric framing narrows responsibility by overlooking both animal suffering and the ecological harm embedded in human activity. Many industries linked to environmental degradation – agriculture, fashion, energy – are also major sites of animal exploitation, yet sustainable development rarely challenges the systems driving these harms (Wadham, 2020). In corporate contexts, sustainability often becomes a managerial tool for

DOI: 10.4324/9781003585541-3

efficiency or reputation rather than ethical reflection. These assumptions are reinforced within business school curricula and NGO discourse, effectively teaching future managers that sustainability can ignore animal exploitation while still appearing progressive. Such approaches ultimately reproduce ecological injustice by rendering animal suffering invisible and reinforcing their commodification (Boscardin & Bossert, 2015).

Given these exclusions, it is striking how little space sustainability debates make for animals, despite extensive evidence of animal sentience and agency (Sebo et al., 2022). This absence shapes which environmental questions are seen as consequential and which remain unasked. Including animals in sustainability thinking broadens the moral and political horizons of the field, challenging the logic that reduces living beings to resources and opening pathways to more inclusive forms of development (Kopnina et al., 2024). Scholars such as Narayanan (2016) and Tsing et al. (2020) argue that animal perspectives deepen our understanding of justice and interdependence amid planetary crisis. This more-than-human view also aligns with pedagogies centred on critical reflection, relational ethics and decentring the human (Kopnina et al., 2025a). Incorporating animals into these frameworks allows sustainability to address exploitation more effectively and imagine caring, thriving futures for all.

Elephants offer a powerful lens through which to reimagine sustainable development. As intelligent, emotionally complex beings they are more than biodiversity symbols or economic assets. They are individuals with unique interests and vulnerabilities (Kopnina, 2020; van de Water et al., 2022). Across Asia and Africa, elephants are revered yet exploited through tourism, logging and conservation economies (Kopnina & Baker, 2023; Terada, 2021). This tension makes them a compelling case for exploring how animals might be meaningfully included in sustainability thinking.

As "charismatic megafauna" (Lorimer, 2015), elephants attract conservation attention partly because of their emotional appeal (Barua et al., 2010). While this can overshadow less visible species, it has generated important work on elephant welfare and human–elephant relations. Elephants also expose the anthropocentric bias of the global sustainability agenda. SDG 15 ("Life on land") focuses on biodiversity protection but frames animals like elephants as conservation targets or ecological units rather than sentient beings with needs and rights (Kopnina, 2020). In practice, they face escalating conflict with humans through deforestation, monoculture and infrastructure expansion (Banerjee et al., 2024; Barua, 2024). These problems point to the limits of technocratic solutions and the need for relational, community-based approaches that support coexistence (Kei et al., 2025; van de Water et al., 2022). Elephants thus serve not as romantic symbols, but as a critical entry point for rethinking sustainability in more ethical and inclusive terms. As such, the chapter provides a useful starting point for the book, by showing how sustainability debates can be rethought through an interspecies lens.

Introducing our expert collaborator for this chapter

Helen Kopnina is a leading critical voice in sustainability studies. Trained in social anthropology at the University of Cambridge, she has focused much of her career on the ethical and ecological implications of sustainability frameworks. She argues that mainstream approaches, especially the SDGs, prioritise human wellbeing and economic growth while sidelining nonhuman animals and ecosystems (Kopnina, 2019).

Her ecocentric perspective positions animals as sentient co-inhabitants rather than resources (Kopnina, 2020). At the same time, Helen's work emphasises that ecocentrism alone is not enough. She repeatedly highlights the productive tensions between valuing ecological collectives (species, habitats, ecosystems) and addressing the moral claims of individual animals. Instead of treating these perspectives as mutually exclusive, she argues for an ethical pluralism that recognises both (Kopnina & Washington, 2020). This means asking not only how to protect ecosystems, but also how human-driven conditions – such as industrial agriculture, population growth, overconsumption and habitat conversion – create false dilemmas in which the interests of individuals and collectives appear to conflict (Kopnina et al., 2024). By foregrounding the human causes of these pressures, she shows that caring for ecosystems and caring for individual animals are not opposing goals but intertwined responsibilities. Her widely cited work on anthropocentrism critiques the invisibility and instrumentalisation of animals and argues for the urgent need to recognise animal agency in global environmental governance.

In our interview, Helen highlighted the contradictions in sustainability discourse: Animals often appear as symbols in conservation and tourism, but are excluded as stakeholders when decisions are made. This exclusion has material (or "real-world") consequences, especially for species who live at the intersection of culture, ecology and economics.

Her work on elephants illustrates these tensions. In Thailand and elsewhere they are celebrated in national mythology, yet exploited within the tourism and logging industries. She uses this example to advocate for an "interspecies approach" to sustainability – one that honours animals' rights and fosters more inclusive forms of development. Through her scholarship and teaching, Helen has helped shape a more ethically robust and ecologically inclusive sustainability agenda, making her a vital voice in this conversation.

Conceptual foundations: What is sustainable development?

Sustainability refers broadly to the ability of ecological, social and economic systems to endure over time. In global policy and governance, however, *sustainable development* has become the dominant framing because it aligns environmental concerns with human-centred goals such as economic growth, poverty reduction and public health (Ruggerio, 2021). Defined in the Brundtland Report (1987) as "development that meets the needs of the present without compromising the ability of future generations to meet their own needs," it offers a policy-friendly vision that enables governments and organisations to address environmental challenges without disrupting dominant economic priorities. Yet this compatibility often results in addressing symptoms in contrast to root causes. Major drivers of biodiversity loss – industrial agriculture, livestock expansion, habitat destruction and broader human ecological impacts – are rarely confronted directly (Wadham, 2020). At the same time, inconsistencies persist in species protection, with some wild animals labelled "invasive" while the far greater ecological impacts of domesticated animals receive limited scrutiny (Kopnina & Coghlan, 2022).

In business and management studies, sustainability is frequently treated as a managerial tool. Research focuses on embedding sustainability into corporate governance, performance metrics and strategy (Sachs et al., 2019). These efforts have generated innovations in ESG reporting and impact assessments, yet they remain oriented towards efficiency and risk management instead of deeper ethical reflection. For instance, the UK government's Dasgupta Review quantifies the economic value of nature to influence policymaking but ultimately reaffirms ecosystems as resources for human benefit (Dasgupta, 2021).

This approach assumes human wellbeing depends on continued economic growth while largely ignoring who bears the costs of achieving it (Bendell, 2022; Selin et al., 2024). Environmental "goods" are treated as substitutable or compensatory, sidestepping the biophysical limits identified by Daly (2019) and others since the 1970s. Consequently, much mainstream (or "weak") sustainability scholarship leaves intact the extractive economic logic driving inequality, environmental degradation and species decline. These patterns echo wider governance critiques showing how institutional structures dilute environmental commitments to preserve existing economic models (Biermann et al., 2022).

One attempt to strengthen sustainability involves integrating biodiversity more explicitly into governance tools. Recent work on biodiversity and extinction accounting demonstrates how ecological concerns can be translated into corporate reporting. Helen and colleagues develop practical extinction-accounting frameworks that track corporate contributions to species decline and ecosystem disruption (Kopnina et al., 2024). Treating biodiversity as a core accountability issue – not a reputational add-on – creates clearer routes for businesses to move beyond greenwashing and take genuine responsibility for ecological harm.

Critical approaches to sustainable development

Alternative frameworks challenge the anthropocentric assumptions that underpin mainstream sustainable development. Raworth's (2018) "doughnut economics" identifies a safe and just space bounded by ecological ceilings and social foundations, while Rockström et al. (2009) warn of planetary boundaries that cannot be exceeded without destabilising Earth systems. The degrowth movement – although internally diverse in motivations and scope – contributes an ecological strand exemplified by scholars like Hickel (2020), who argue that sustainability is fundamentally incompatible with continued economic expansion in wealthy nations. Hickel critiques development's colonial legacy and calls for redistributive, post-growth futures centred on ecological balance.

Though still centred on human wellbeing, such "strong sustainability" approaches stress that ecological limits are non-negotiable. By contrast, Helen argues that the familiar "three pillars" model misleadingly suggests equivalence between economic, social and environmental objectives, when in practice ecological concerns are routinely subordinated to economic priorities (Kopnina, 2020). This contributes to accelerating biodiversity loss. Wildlife populations have declined by 73% since 1970 (WWF, 2024), and Bar-On et al. (2018) show that whereas wild mammals once dominated global biomass, they now represent only a tiny fraction. Humans and domesticated animals together make up around 96% of all terrestrial mammalian biomass.

This disparity exposes a deeper inconsistency in sustainability discourse, such as in debates about "invasive species." Wild animals are often blamed for ecosystem disruption, while the far-reaching impacts of domesticated species – driven by human land use, feed production and industrial agriculture – are largely exempt from such labels. This framing shifts responsibility away from the systems causing biodiversity decline and unfairly positions wildlife as the primary threat (Kopnina & Coghlan, 2022).

These ecological crises have prompted not only scientific warnings but also contributions from Indigenous and post-development scholars who argue that sustainability cannot be understood without recognising histories of colonialism, dispossession and alternative ways of relating to the nonhuman world (Whyte, 2020). Simpson (2017) advocates "grounded normativity," a mode of governance rooted in relational ethics, kinship and place rather than abstract metrics. Poelina et al. (2019), for example, offer a water-governance model that treats rivers as ancestral beings, not as "natural capital." Gudynas (2011) draws on the Latin American concept of buen vivir ("good living"), emphasising reciprocity, ecological balance and the embeddedness of human communities in nature. Correspondingly, Escobar (2020) argues for "pluriversal" politics that reject universal development models and instead make space for diverse worldviews, including non-Western ontologies that foreground interdependence with animals and ecosystems.

Across these frameworks, several shared principles emerge, namely decentring the human, challenging growth dependency and resisting the reduction of life to economic value. They compel us to ask not only *how* to sustain life but *whose* lives are protected – and according to whose knowledge systems. This is especially relevant for animals, who remain largely invisible within mainstream sustainability discourse. For sustainable development to be meaningful, it must address the ethical implications of species loss, centre ecological integrity, and recognise interdependence across species. Without such a reorientation, sustainability risks becoming a hollow technocratic exercise, disconnected from the communities – human and nonhuman – it purports to protect.

Towards an interspecies approach to sustainable development

As we have outlined above, animals are routinely excluded from global sustainable development discourse. This omission represents a form of structural anthropocentrism, where human priorities remain paramount and nonhuman animals are considered only insofar as they serve economic or environmental functions. Their interests are rarely considered in their own right. This section explores how acknowledging animals – as stakeholders, ethical agents and kin – might transform the ethical foundations and practical structures of sustainability.

Animals as stakeholders within sustainable debates

Discussions of sustainability rarely consider animals as meaningful stakeholders, even though many policy agendas – such as the SDGs – directly affect their lives and often constrain their wellbeing (Sebo et al., 2022). One way to begin challenging this anthropocentric bias is by including animals within the language and logic of one of sustainability's most entrenched frameworks, that of stakeholder theory. Together with her co-authors, Helen has argued precisely for this shift, advocating for the recognition of animals as legitimate stakeholders whose interests and agency must be included in sustainability decision-making processes (Kopnina, 2019; Kopnina & Washington, 2020). Her work on elephants offers a concrete, well-researched case through which to explore how animals live at the intersection of ecology, economy and ethics – and how they might be repositioned as stakeholders in sustainability debates.

In contexts such as Thailand's tourism industry, elephants can function as cultural icons and economic assets while still being subjected to harsh training regimes, confinement and forced labour (Kopnina & Baker, 2023). In such settings, the elephant becomes both symbol and commodity – revered yet subordinated. Importantly, this symbolism is not universal. Relationships with elephants differ across cultures and regions, underscoring the need for context-specific analysis as opposed to generalised claims. Treating elephants as stakeholders therefore requires attending to their agency and wellbeing within particular socio-ecological settings, in preference to reducing them to their human-assigned roles.

This framing shifts the question from "how do we protect animals?" to "how do we include animals in sustainability itself?" If animals are part of our shared socio-ecological systems, then their interests, needs and vulnerabilities must be accounted for within the ethical architecture of sustainable development. This challenges business-as-usual approaches in sustainability education, which tend to treat animals as externalities not ethical agents (Kopnina et al., 2025a). For instance, in the case of elephant "stakeholders" in tourism or conservation, treating them as ethical agents would require not just regulating their use, but rethinking the very systems that commodify their lives for human benefit.

However, this move is not without its tensions. As Buller et al. (2018) note, stakeholder frameworks risk anthropomorphising animals – attributing human-like motivations and preferences to them – in ways that can obscure their distinct forms of agency. For elephants, this might mean assuming they "want" to participate in tourism or conservation simply because they display behaviours interpreted as cooperation, rather than recognising these as responses to coercion or stress. There is also the danger of instrumentalising their inclusion – treating elephants as symbolic stakeholders to justify human-centred decisions under the guise of inclusivity, such as showcasing captive elephants in corporate sustainability campaigns while continuing exploitative practices.

Similarly, Wallach et al. (2020) caution that even well-intentioned stakeholder approaches can reinforce managerial logics. In practice, this might take the form of conservation projects where elephants are microchipped, tracked and fenced – not for their wellbeing, but to serve tourism goals or prevent human/wildlife conflict in ways that prioritise human interests. Here, "care" becomes control, and protection can translate into surveillance or confinement. Still, many scholars see value in the stakeholder model. Nista et al. (2020), for instance, argue that recognising animals as stakeholders can bring ethical considerations into land-use and biodiversity policy. For elephants, this might involve safeguarding migratory corridors not only for ecosystem function but for the wellbeing of the elephants who rely on them.

Rethinking participation and ethics

True interspecies sustainability demands a radical rethink of ethics and participation. It is not simply about extending existing rights frameworks to animals but about interrogating the structures that define who counts, how, and why. Ecocentric ethics, as advanced by Helen and others (e.g. Kopnina et al., 2022) suggest that ethical consideration should not be limited to humans or even to sentient beings alone, but extend to whole ecosystems as living entities with moral claims. For example, adopting an ecocentric perspective in elephant conservation would mean valuing elephants not merely for their cultural or economic significance, but as integral members of ecosystems whose needs and relationships shape the health of the environment as a whole.

Yet as Wallach et al. (2020) argue in their work on compassionate conservation, this ecocentrism should not come at the cost of individual animals. They propose a relational ethic that values both the individual and the collective, the sentient and the systemic – a stance that requires humility, nuance, and care. Their "four tenets" of compassionate conservation – do no harm, individuals matter, inclusivity, and peaceful coexistence – are highly relevant for sustainability thinking. Applied to elephants, this would mean rejecting practices such as violent training or forced captivity in tourism, and instead fostering environments where elephants can live freely and safely as both ecological participants and sentient individuals.

Philosophically, this aligns with Narayanan's (2016) call for a model of "sociozoological justice," in which animals are not left to "fall between the cracks of 'not-quite-human' and 'not-quite-nature'" but seen as ethical and political subjects. This requires a shift from stewardship to kinship – from managing animals to relating with them in ways grounded in care and respect. For elephants, this would mean abandoning their treatment as resources for tourism or conservation and recognising them as kin: emotionally intelligent, socially complex beings whose wellbeing is intimately bound up with human choices.

Realising such a shift requires policy change as well as ethical reflection. Visseren-Hamakers (2020) proposes the creation of an 18th SDG to focus explicitly on animals. She argues that while the existing 17 goals affect animals indirectly, they fail to account for animals as subjects of justice. Simply adding an additional goal could elevate animal welfare and agency within sustainability governance. However, such a move also risks reinforcing the marginalisation of animals – implying they are irrelevant to the rest of the framework. Tallberg and Tienari (2022) go further, calling for a posthuman affirmative ethics that reconfigures responsibility and voice within global agendas. They argue that recognising animals as agents with interests means not just giving them a separate "box," but embedding interspecies ethics across all levels of governance and practice.

In this light, the value of an 18th SDG lies less in compartmentalising animals and more in catalysing a shift towards integrated, relational thinking. For such a proposal to advance sociozoological justice, it must act not as a policy add-on, but as a catalyst for reforming the SDG framework as a whole.

Navigating the tensions between ecocentrism and compassion

While animal welfare remains a vital concern, it can sometimes serve as a limit instead of a point of departure for deeper ethical reflection. Helen's work highlights a core tension between ecocentric approaches (which prioritise ecological wholes, such as species or ecosystems) and animal rights/welfare approaches (which focus on the experience of individual animals (Kopnina et al., 2022)). These tensions are especially pronounced in conservation contexts, where protecting biodiversity may come at the expense of individuals – via culling, relocation or captivity. This stance challenges conservation paradigms that permit harm to individuals for the "greater good" – a utilitarianism that often excuses violence in the name of biodiversity. For instance, efforts to manage elephant populations in fragmented habitats can lead to practices such as darting, forced translocation or confinement, often justified in the name of ecological balance but traumatic for the animals involved. Similar dilemmas arise with wild ponies in the UK and Australia, for example, where culling is framed as care *for* herd health, even as it disregards care *about* individual lives, revealing how these ethical trade-offs cut across species (Castelló & Santiago-Ávila, 2022; Wadham, 2020).

This is not an abstract debate but a live ethical problem with real material consequences for those involved. Buller et al. (2018) and van de Water et al. (2022) document how elephants in captivity are often kept under conditions that would be deemed unacceptable for any other sentient being, despite being justified on grounds of conservation, tradition or tourism. The result is a bifurcated ethical system in which some animals are sacrificed for systemic goals while others are venerated. Yet this need not be an either/or. Helen argues for ethical pluralism – an approach that allows for both ecological and sentientist commitments (Kopnina et al., 2022). This means rethinking what "wellbeing" means in sustainability and asking whether our frameworks are truly inclusive. For example, Kei et al. (2025) advocate for community-based coexistence models that support both human and animal needs, grounded in local ethics and mutual respect. Ultimately, navigating these tensions requires an ethic of humility – an acknowledgement that our frameworks are incomplete, our knowledge partial and our responsibilities profound.

From conservation to coexistence

Barua (2024) describes elephants as "charismatic mediators" in plantation landscapes, navigating – and suffering from – the pressures of agricultural expansion. Terada (2021) shows how conservation practices can reproduce colonial dynamics, imposing external models on Indigenous lands. In response, Canney (2021) proposes "reconciliation ecology," redesigning human landscapes to foster coexistence not exclusion.

Ethical models of elephant rehabilitation and tourism – such as those described by van de Water et al. (2022) – demonstrate alternatives, but remain the exception. As Kei et al. (2025) argue, what's needed is not just technical reform but moral transformation: participatory, interspecies models that centre both human and elephant needs. Helen similarly urges a move from "conservation" to "coexistence," where elephants are respected not as assets, but as neighbours (Kopnina & Baker, 2023).

Incorporating elephants more actively into sustainability debates reveals that animals are not simply victims of development – they are also teachers and guides. Sustainability must be reclaimed not as a managerial tool, but as a moral, relational practice rooted in mutual care. It's not about adding animals into existing systems – it's about transforming the systems themselves.

Directions for future research

Future research must confront the conceptual and practical gaps in sustainability thinking. Helen's body of work suggests that the very foundations of sustainable development warrant critical re-evaluation. Rather than simply expanding existing frameworks to include animals, then, scholars should interrogate the anthropocentric assumptions that underpin them. This includes challenging the instrumental treatment of animals in the SDGs – as biodiversity "resources" or providers of ecosystem "services" – and instead recognising them as sentient, relational beings with intrinsic worth and political relevance (Kopnina, 2020; Kopnina et al., 2025a).

One key area for future inquiry is how interspecies ethics might inform sustainability governance. Visseren-Hamakers (2020) has proposed an 18th SDG focused explicitly on animals. While this could help surface nonhuman concerns within global policy, it risks sidelining them further by treating them as separate from mainstream goals. Research should therefore assess both the potential and pitfalls of such a proposal, while exploring ways to integrate interspecies justice across the entire SDG framework. This could include empirical studies of how development projects – particularly in infrastructure, tourism and conservation – affect animal lives, and how governance models might better support coexistence, kinship and care-based ethics (Kei et al., 2025; Wallach et al., 2020).

Second, scholars should continue to explore the pedagogical dimensions of interspecies sustainability. Helen calls for a shift in sustainability teaching towards ecocentric and interspecies pedagogy (Kopnina et al., 2025b). This involves equipping students with the tools to critically engage with growth-based paradigms, fostering ethical literacy around nonhuman lives. Future work could build on emerging models like ecopedagogy or "compassionate conservation" to create new curricula, case studies and learning methods that foreground interspecies concerns.

Third, theoretical contributions are also needed to advance post-anthropocentric ethics. Frigo et al. (2024) argue for moving beyond "shallow geoethics" by embracing ecocentric values. Future studies could develop frameworks that centre animal agency and nonhuman interests in political and moral decision-making. This might involve legal innovations like animal personhood, or deeper engagement with indigenous ontologies that reject human exceptionalism. Such work would be especially timely in the lead-up to global forums like the UN Summit of the Future, or in shaping the policy landscape that will succeed Agenda 2030.

Finally, future research must be intersectional and interdisciplinary. Interspecies injustice intersects with colonialism, capitalism and global inequality (Dietz & York, 2015). Elephants, for example, are ecologically significant, culturally symbolic and deeply entangled in political economies of tourism, conservation and land use. Their case can guide studies that examine animal experiences across species, regions and governance regimes – especially beyond the Global North and beyond charismatic wildlife.

In short, the future of sustainability research lies in decentring the human, foregrounding animal lives and engaging actively with the ethical complexity of shared existence. Only by embracing these challenges can sustainable development evolve into a more inclusive, just and ecologically grounded project.

Ethical implications

Including animals in sustainability frameworks raises complex ethical and political questions that demand careful navigation. Efforts to centre animals must avoid repeating the injustices of past conservation models, which have too often displaced marginalised communities or imposed rigid, top-down structures (Kei et al., 2025; Wallach et al., 2020). A truly interspecies ethics must be intersectional – one that seeks justice for both human and nonhuman lives, and recognises that protecting one need not come at the cost of the other.

At the same time, species differences matter. As Narayanan (2016) argues, we must resist one-size-fits-all approaches that ignore the diversity of animal lives. Elephants are not the same as clownfish or beetles, and their needs, relationships and forms of agency differ. Ethical inclusion means recognising this diversity while affirming a shared right to exist on their own terms. Wallach et al.'s (2020) appeal for ethical pluralism reflects this need for nuance, suggesting that conservation can support both ecological health and individual wellbeing without reducing one to the other.

Finally, it is heartening to note that public concern for animals is often ahead of political action (Monterrubio et al., 2024). In many cases, people intuitively reject animal exploitation long before policies catch up. This public sentiment offers an opportunity: Not to simply bolt animals onto existing sustainability agendas, but to reimagine those agendas through a more relational, inclusive lens. Proposals like the 18th SDG (Visseren-Hamakers, 2020) or the other policy adaptations suggested by Tallberg and Tienari (2022) should be seen not as endpoints, but as openings for deeper structural transformation.

Practical implications for responsible management

An interspecies approach to sustainable development urges business leaders to rethink their responsibilities through a broader ethical lens. Rather than viewing sustainability solely in terms of human welfare or economic growth, responsible managers must acknowledge that business operations also significantly impact animals and ecosystems. This shift involves moving from a narrow, anthropocentric model towards one that centres justice and care across species (Kopnina, 2020).

For managers, this means re-evaluating key operational areas. Supply chains, for example, often overlook animal welfare, despite the fact that sectors such as food, fashion, tourism and pharmaceuticals rely directly on animal bodies, labour or habitats. Companies can adopt animal-inclusive auditing tools to identify where harm occurs and pursue more ethical alternatives. Helen's work on elephant tourism in Thailand highlights how operators can move away from exploitative practices – like elephant riding – and instead support observation-based models rooted in rehabilitation and non-interference. These initiatives not only enhance animal welfare but also foster community partnerships and long-term sustainability.

In food production, leaders can support plant-based innovation or regenerative systems that reduce harm to both animals and ecosystems. Similarly, firms in the beauty or biomedical industries can prioritise cruelty-free testing and alternatives that are already widely available. These changes align with growing consumer demand for ethical transparency and offer reputational benefits to early adopters.

More fundamentally, responsible management should involve institutionalising interspecies ethics. An expanded interspecies version of stakeholder theory opens up the possibility of formally recognising animals within governance structures (Kopnina & Washington, 2020; Tallberg et al., 2022). This could involve assigning internal advocates, ethics officers, or creating advisory panels with representation from animal welfare experts. Some businesses have explored appointing "animal proxies" or guardians to ensure nonhuman interests are included in strategic decision-making – similar to how some companies now consider future generations or ecosystems in their planning (Frigo et al., 2024; Meijer, 2025).

Partnering with animal advocacy organisations, conservation networks or sanctuaries can further guide companies through complex ethical terrain. These collaborations can support evidence-based decision-making and enhance public trust – especially as consumer expectations evolve (Tallberg et al., 2022).

This approach does not demand instant or radical transformation. Rather, it invites managers to ask deeper questions: How does our business affect animal lives? How can we reduce harm or promote coexistence? And what ethical values should our organisation prioritise in an age of ecological crisis?

Ultimately, interspecies responsibility expands the moral horizons of business. As Frigo et al. (2024) suggest, true ethical transformation in management requires more than shallow sustainability, it demands a commitment to care, reciprocity and ecological humility. By taking animals seriously, businesses can model a more inclusive and sustainable form of leadership fit for the challenges of the 21st century. Taken together, these arguments demonstrate how rethinking sustainable development through animals reshapes not only environmental policy, but the ethical foundations of business itself.

Reflective activities inside/outside the classroom

A Revision questions that test knowledge ("what?")

- A1 What are the Sustainable Development Goals (SDGs) and who created them?
- A2 What role do elephants play in tourism in Thailand?
- A3 What other industries rely on animals in their operations?
- A4 What is the purpose of an 18th SDG proposed by some researchers?

B Reflective questions that build understanding ("so what?")

- B1 Why do you think animals are often left out of sustainability plans?
- B2 How does the treatment of animals in industries such as tourism or farming challenge dominant sustainability narratives?
- B3 What are the potential risks of symbolic inclusion (e.g. featuring animals in sustainability branding without changing harmful practices)?
- B4 Why should we involve animals in corporate governance and how might that change the way decisions are made?

C Take action ("do")

- C1 Choose an everyday industry (e.g. food, tourism, fashion, cosmetics) and explore how it affects animals. Make a simple chart or diagram showing where animals are involved and what kinds of impacts (positive or negative) they experience. Then, reflect on what a more sustainable, animal-friendly version of this industry might look like.
- C2 Select one animal mentioned in the book (e.g. elephants, fish, bees) and design a short teaching tool (poster, TikTok or slide) to explain how this species connects to sustainability. Include facts about their lives, why they are important for ecosystems and the environment and what kinds of support they need.
- C3 Write a short "sustainability pledge" for a business that includes animals. List three concrete actions the company could take (e.g. cruelty-free testing, ethical sourcing, partnerships with sanctuaries), and identify two possible barriers to making these changes.
- C4 Watch a short video or explore a website about an animal welfare or conservation project (e.g. an elephant sanctuary in Thailand). Write 200 words reflecting on how the project links to sustainable development. What values does it promote? Could it go further in centring animal needs?

D Ideas to discuss with colleagues, friends and family ("share")

- D1 Have you ever participated in animal-based tourism (e.g. riding elephants or camels, visiting zoos or aquariums)? What messages were you given about the animals? Would you respond differently now?
- D2 Share a recent advert, social media post or brand campaign that used animals in its messaging. How were the animals portrayed? What assumptions or emotions were being used?
- D3 Talk with friends or family about how different cultural backgrounds shape views on animals' roles in society. For example, are elephants seen as sacred, useful, dangerous – or something else?
- D4 What would it mean to take animal perspectives seriously in your own workplace or future career? How could animals be acknowledged or protected in your professional decisions?

Key ideas

Sustainability vs. sustainable development

Sustainability is the goal, or the long-term endurance of ecological, social and economic systems. Sustainable development is the path towards that goal; a policy-driven framework that aligns environmental concerns with human priorities like better health and poverty reduction. But it is inherently contradictory, often prioritising short-term economic gains over the environmental foundations that make sustainability possible (Ruggerio, 2021).

Ecocentrism

The idea that nature – including animals, plants and ecosystems – has value on its own terms, not just for the benefits that it brings to humans. It asks us to care about the whole Earth, not just people, and supports bigger changes to protect the planet, centring on ecological justice and systemic change (Frigo et al., 2024; Kopnina et al., 2022).

Interspecies justice

Treating animals and ecosystems as part of our moral and political responsibilities – not just humans. It asks us to respect animal needs, rights and roles in society, and to include them in decisions about sustainability and the future (Kopnina & Washington, 2020; Tallberg & Tienari, 2022).

Biodiversity

Refers to the variety of life on Earth, including genetic, species, and ecosystem diversity. While commonly used in sustainability discourse, it's often framed instrumentally. Critical scholars call for recognising biodiversity as more than a resource – highlighting its ethical and relational significance (Kopnina, 2019).

Compassionate conservation

An approach that prioritises the welfare of individual animals alongside broader ecological goals. Its key principles – do no harm, individuals matter, inclusivity and peaceful coexistence – challenge practices like culling and captivity in the name of conservation (van de Water et al., 2022; Wallach et al., 2020).

Sentience

The capacity to feel things like pain, pleasure or fear. When we recognise animals as sentient, we stop seeing them as just resources and start treating them as beings with their own needs. This supports including animals in laws, education and business decisions (Sebo et al., 2022).

References

Banerjee, S., Nayak, D., & Sinha, A. (2024). Adivasi (Tea Tribe) worldviews of living close to wild Asian elephants in Assam, India. *Conservation Biology*, 38(6), e14397. doi:10.1111/cobi.14397.

Bar-On, Y. M., Phillips, R., & Milo, R. (2018). The biomass distribution on Earth. *Proceedings of the National Academy of Sciences*, 115(25), 6506–6511. doi:10.1073/pnas.1711842115.

Barua, M. (2024). *Plantation worlds*. Duke University Press. doi:10.1353/book.126549.

Barua, M., Tamuly, J., & Ahmed, R. A. (2010). Mutiny or clear sailing? Examining the role of the Asian elephant as a flagship species. *Human Dimensions of Wildlife*, 15 (2), 145–160. doi:10.1080/10871200903536176.

Bendell, J. (2022). Replacing sustainable development: Potential frameworks for international cooperation in an era of increasing crises and disasters. *Sustainability*, 14(13), 8185. doi:10.3390/su14138185.

Biermann, F., Hickmann, T., Sénit, C. A., Beisheim, M., Bernstein, S., Chasek, P., ... & Wicke, B. (2022). Scientific evidence on the political impact of the Sustainable Development Goals. *Nature Sustainability*, 5(9), 795–800. doi:10.1038/s41893-022-00909-5.

Boscardin, L., & Bossert, L. (2015). Sustainable development and nonhuman animals: Why anthropocentric concepts of sustainability are outdated and need to be extended. In S. Meisch, J. Lundershausen, L. Bossert, & M. Rockoff (Eds), *Ethics of science in the research for sustainable development* (pp. 323–352). Nomos Verlagsgesellschaft. doi:10.5771/9783845258430-323.

Brundtland, G. H. (1987). *Our common future: The World Commission on Environment and Development*. Oxford University Press.

Buller, H., Blokhuis, H., Jensen, P., & Keeling, L. (2018). Towards farm animal welfare and sustainability. *Animals*, 8(6), 81. doi:10.3390/ani8060081.

Canney, S. M. (2021). Making space for nature: Elephant conservation in Mali as a case study in sustainability. *Environment: Science and Policy for Sustainable Development*, 63(2), 4–15. doi:10.1080/00139157.2021.1871292.

Castelló, P. P., & Santiago-Ávila, F. J. (2022). Conservation after sovereignty: Deconstructing Australian policies against horses with a plea and proposal. *Hypatia*, 37 (1), 136–163. doi:10.1017/hyp. 2021.72.

Daly, H. (2019). Growthism: Its ecological, economic and ethical limits. *Real-World Economics Review*, 87, 9–22.

Dasgupta, P. (2021). *The Economics of Biodiversity: The Dasgupta Review. Abridged Version*. HM Treasury.

Dietz, T., & York, R. (2015). Animals, capital and sustainability. *Human Ecology Review*, 22(1), 35–53. doi:10.22459/HER.22.01.2015.03.

Escobar, A. (2020). *Pluriversal politics: The real and the possible*. Duke University Press. doi:10.1515/9781478012108.

Frigo, G., Ifanger, L. A., Greco, R., Kopnina, H., & Hillerbrand, R. (2024). Shallow vs. deep geoethics: Moving beyond anthropocentric views. *Journal of Agricultural and Environmental Ethics*, 37(1), 2. doi:10.1007/s10806-023-09920-y.

Gudynas, E. (2011). Buen Vivir: Today's tomorrow. *Development*, 54(4), 441–447. doi:10.1057/dev.2011.86.

Hickel, J. (2020). *Less is more: How degrowth will save the world*. Random House.

Kei, L. H., Bakri, N. A. A. A. B., Lin, P. Q., Khairunizan, H. K. B., Yuan, K. H. Q., & Abdullah, N. B. M. (2025). Mitigating human–elephant conflict: A comprehensive review of strategies for sustainable coexistence in support of SDG 15. *Journal of Lifestyle and SDGs Review*, 5(2), e2756–e2756. doi:10.47172/2965-730X.SDGsReview.v5.n02.pe02756.

Kopnina, H. (2019). Anthropocentrism: Problem of human-centred ethics in Sustainable Development Goals. In W. Leal Filho, A. Azul, L. Brandli, P. Özuyar, & T. Wall (Eds), *Life on land: Encyclopedia of the UN Sustainable Development Goals* (pp. 1–9). Springer. doi:10.1007/978-3-319-71065-5_105-1.

Kopnina, H. (2020). Of elephants and men. *Animal Sentience*, 5(28), 2. doi:10.51291/2377-7478.1553.

Kopnina, H., & Baker, L. (2023). Conservation, animal wellbeing, and Indigenous participation at an elephant sanctuary in Mondulkiri, Cambodia. *Society & Animals*, 32(5–6), 560–581. doi:10.1163/15685306-bja10126.

Kopnina, H., & Coghlan, S. (2022). Invasion biology and its discontents: Human supremacy, language, and animal treatment. *Visions for Sustainability, 18*, 1–28. doi:10.13135/2384-8677/6512.

Kopnina, H., & Washington, H. (2020). *Conservation: Integrating social and ecological justice*. Springer. doi:10.1007/978-3-030-13905-6_1.

Kopnina, H., Gray, J., Lynn, W., Heister, A., & Srivastava, R. (2022). Uniting ecocentric and animal ethics: Combining non-anthropocentric approaches in conservation and the care of domestic animals. *Ethics, Policy & Environment*, 26(2), 265–286. doi:10.1080/21550085.2022.2127295.

Kopnina, H., Zhang, S. R., Anthony, S., Hassan, A., & Maroun, W. (2024). The inclusion of biodiversity into Environmental, Social, and Governance (ESG) framework: A strategic integration of ecocentric extinction accounting. *Journal of Environmental Management*, 351, 119808. doi:10.1016/j.jenvman.2023.119808.

Kopnina, H., Roca, T., & Sorour, M. K. (2025a). Integrating critical thinking about Sustainable Development Goals: Making biodiversity matter for business students. *Journal of Education for Sustainable Development*. doi:10.1177/09734082251336403.

Kopnina, H., Wong, R., Black, K., & Đurović, M. (2025b). Anthropocentrism and climate change: Radical reorientation away from greenwashing and toward degrowth in business education. In S. Brechin & S. Lee (Eds), *Routledge handbook of climate change and society* (pp. 193–207). Routledge.

Lorimer, J. (2015). *Wildlife in the Anthropocene: Conservation after nature*. University of Minnesota Press. doi:10.5749/minnesota/9780816681075.001.0001.

Meijer, E. (2025). *Multispecies dialogues: Doing philosophy with animals, children, the sea and others*. Routledge.

Monterrubio, C., Dashper, K., & Wadham, H. (2024). Contested intangible heritage: Equestrian sport and animal welfare in Mexico. *International Journal of Heritage Studies*, 30(3), 332–348. doi:10.1080/13527258.2023.2294761.

Narayanan, Y. (2016). Where are the animals in sustainable development? Religion and the case for ethical stewardship in animal husbandry. *Sustainable Development*, 24(3), 172–180. doi:10.1002/sd.1619.

Nista, N. A., Jannuzzi, C. A. S. C., Falsarella, O. M., & Benedicto, S. C. D. (2020). Society and sustainable development: Animal rights in sustainability discourse. *Ambiente & Sociedade*, 23, e02782. doi:10.1590/1809-4422asoc20180278r2vu2020l4ao.

Poelina, A., Taylor, K. S., & Perdrisat, I. (2019). Martuwarra Fitzroy River Council: An Indigenous cultural approach to collaborative water governance. *Australasian Journal of Environmental Management*, 26(3), 236–254. doi:10.1080/14486563.2019.1651226.

Raworth, K. (2018). *Doughnut economics: Seven ways to think like a 21st century economist*. Chelsea Green Publishing.

Rockström, J., Steffen, W., Noone, K., Persson, Å., Chapin, F. S., Lambin, E. F., ... & Foley, J. A. (2009). A safe operating space for humanity. *Nature*, 461(7263), 472–475. doi:10.1038/461472a.

Ruggerio, C. A. (2021). Sustainability and sustainable development: A review of principles and definitions. *Science of the Total Environment*, 786, 147481. doi:10.1016/j.scitotenv.2021.147481.

Sachs, J. D., Schmidt-Traub, G., Mazzucato, M., Messner, D., Nakicenović, N., & Rockström, J. (2019). Six transformations to achieve the Sustainable Development Goals. *Nature Sustainability*, 2(9), 805–814.

Sebo, J., Verkuijl, C., José Hötzel, M., Achakulwisut, P., Bastos Lima, M., & Green, J. (2022). Sustainable development matters for animals too: Governments have a responsibility to recognize that. *CABI One Health*.

Selin, N. E., Giang, A., & Clark, W. C. (2024). Showcasing advances and building community in modeling for sustainability. *Proceedings of the National Academy of Sciences*, 121(29), e2215689121. doi:10.1073/pnas.2215689121.

Simpson, L. B. (2017). *As we have always done: Indigenous freedom through radical resistance*. University of Minnesota Press. doi:10.5749/j.ctt1pwt77c.

Tallberg, L., & Tienari, J. (2022). Human and nonhuman animals in a posthuman reality: Accreditation schemes as voice? In M. Sandberg & J. Tienari (Eds), *Transformative action for sustainable outcomes* (pp. 127–133). Routledge. doi:10.4324/9781003229728-23.

Tallberg, L., García-Rosell, J. C., & Haanpää, M. (2022). Human–animal relations in business and society: Advancing the feminist interpretation of stakeholder theory. *Journal of Business Ethics*, 180(1), 1–16.

Terada, S. (2021). Building human–elephant relationships based on science and local ownership: A long-lasting issue in the era of Sustainable Development Goals. *Animal Conservation*, 24(5), 738–739.

Tsing, A. L., Deger, J., Keleman Saxena, A., & Zhou, F. (2020). *Feral atlas: The more-than-human Anthropocene*. Stanford University Press. doi:10.21627/2020fa.
van de Water, A., Di Minin, E., & Slotow, R. (2022). Human–elephant coexistence through aligning conservation with societal aspirations. *Global Ecology and Conservation*, 37, e02165. doi:10.1016/j.gecco.2022.e02165.
Visseren-Hamakers, I. J. (2020). The 18th Sustainable Development Goal. *Earth System Governance*, 3, 100047. doi:10.1016/j.esg.2020.100047.
Wadham, H. (2020). Horse matters: Re-examining sustainability through human–domestic animal relationships. *Sociologia Ruralis*, 60(3), 530–550. doi:10.1111/soru.12293.
Wallach, A. D., Batavia, C., Bekoff, M., Alexander, S., Baker, L., Ben-Ami, D., ... & Ramp, D. (2020). Recognising animal personhood in compassionate conservation. *Conservation Biology*, 34(5), 1097–1106. doi:10.1111/cobi.13494.
Whyte, K. (2020). Indigenous environmental justice: Anti-colonial action through kinship. In B. Coolsaet (Ed.), *Environmental justice* (pp. 266–278). Routledge. doi:10.4324/9780429029585-2.
WWF. (2024). *Living planet report 2024: Tracking the health of nature*. World Wide Fund for Nature.

2

ANIMALS AND WELFARE

Introduction

Contemporary livestock systems are organised through business and management logics of efficiency, risk management, quality assurance and reputational governance within which animal health and welfare function as strategic economic concerns. Veterinary expertise is embedded in supply chains to stabilise production, secure food safety, manage biological risk and position animal bodies as sites of value creation (Amos & Sullivan, 2018; Enticott et al., 2011; Hinchliffe et al., 2016). Drawing on Lewis Holloway's work on agri-food systems, together with feminist and posthumanist theories of care, this chapter reimagines livestock not as passive resources but as active participants in interspecies organisations and networks of care.

The arrangements outlined above have deep historical roots. Veterinary medicine can be traced back to Mesopotamia around 3000 BC and was institutionalised in Europe with the founding of the first veterinary school in France in 1761, followed by the London Veterinary College in 1791. Early practice focused on equine medicine linked to military and transport needs, before expanding to cattle and other livestock as agricultural economies intensified. The industrialisation of animal agriculture in the 19th and 20th centuries dramatically increased productivity and human wellbeing, embedding veterinary practice within systems of labour organisation, economic coordination and market expansion (Yeates, 2013).

These business-driven transformations also produced ecological and ethical consequences. Despite scientific and managerial advances, endemic disease and welfare problems persist, exposing biological, organisational and moral contradictions within supply-chain-based systems of performance governance. Animal health is now not only a veterinary issue but a

DOI: 10.4324/9781003585541-4

reputational and managerial one. Corporate retailers, processors and certification bodies increasingly frame welfare as an indicator of quality, safety and ethics via corporate social responsibility (CSR) strategies that signal business responsibility for animals in constrained commercial environments (Amos & Sullivan, 2018; Janssens, 2021; Köllen & Schneeberger, 2023; Lever & Evans, 2017). For this reason, welfare is not a peripheral ethical concern but a central organising principle within contemporary agri-food systems. This chapter therefore reframes livestock welfare not as a matter of compliance, but as a question of interspecies governance and, ultimately, of what responsible management means in systems structured around animal life and death.

Veterinary medicine as a management regime

Swabe (1999) reframes veterinary medicine as a social rather than purely technical process, introducing the concept of "veterinary regime" to describe institutionalised systems that safeguard human interests by regulating animal health. These regimes reflect long-term transformations in human–animal relations and exemplify what Foucault (2007) termed biopolitics: The governance of life through surveillance, optimisation and control. Welfare is defined through human priorities such as productivity and biosecurity, aligned with managerial logics of standardisation, audit and risk (Hinchliffe et al., 2016).

Production diseases exemplify the consequences of this approach. Often chronic rather than epidemic, they arise from the organisational and material logics of intensive farming and have become normalised within it (Bellet et al., 2021). Many conditions stem from stress, productivity demands and restricted movement. Their persistence exposes a system that rhetorically promotes welfare while relying on practices that undermine it. Farmers operate under economic and regulatory pressures that constrain the uptake of veterinary advice (Shortall & Calo, 2021). Endemic ill-health is thus backgrounded as animal bodies become sites of "optimisation" rather than partners in care (Buller & Morris, 2003; Hinchliffe et al., 2016).

Within this infrastructure, wellbeing is rendered measurable through data and performance indicators, translating ethical concern into managerial calculation (Buller et al., 2020; Norwood & Lusk, 2011). Disease responses prioritise symptom management or genetic and embodied resilience over ecological or relational health (Bellet et al., 2021). Welfare is therefore stabilised as a management regime – a system of metrics, compliance and risk control that secures food systems and human health while marginalising interspecies wellbeing (Cox, 2022; Swabe, 1999).

Care as responsible management

In recent times, corporate businesses have started to view animal welfare as a source of competitive advantage (Lever & Evans, 2017), but there is little research on the responsibilities of companies towards animals (Janssens, 2021). Feminist and posthumanist scholarship offers a useful way of thinking about the implications of the prevailing anthropocentric worldview. Puig de la Bellacasa (2017) conceptualises care as an ethical and ontological practice of world-making, while Tronto (2020) frames care as political, requiring obligation, skill and responsibility. Mol et al. (2010) describe good practice as tinkering: the situated adjustment of techniques in response to others' needs. Applied to farming, care becomes an embodied, relational practice unfolding within organisational routines and economic constraints. From this perspective, production diseases are not simply biological failures but relational symptoms of stressed multispecies systems.

Addressing them requires shifting from repairing bodies to reconfiguring the relations that produce illness. Veterinary medicine can become a practice of *response-ability* (Haraway, 2008), attentive to animals' capacities to respond, co-adapt and shape care within managed systems. Care thus becomes a governance practice that aligns organisational processes with ethical obligations across species (Coulter, 2016; Donaldson & Kymlicka, 2011). It is not a universal moral category but a situated, affective and political form of world-making. However, as Giraud (2019) and Haraway (2016) warn, care in livestock systems is never innocent. It is entangled with histories of breeding, conditioning and killing that determine which lives are valued. Animals are nurtured yet rendered killable, their deaths rationalised within economic systems translating life into exchange value and what is often termed or thought of as responsible business practice (Lever & Evans, 2017). Recognising this ambivalence does not undermine ethics; it deepens responsibility by situating welfare within the realities of farming life and organisational management.

Farms as interspecies organisations

Livestock farming lies at the intersection of economy, ecology and emotion, where affection and exploitation, care and killing, are inseparable. An animal-centred approach reconceptualises farmed animals not as passive resources but as co-actors in interspecies organisational networks of health. Governance becomes a process of negotiation rather than unilateral control. Van Dam et al. (2023) illustrate this through livestock health governance in southern Africa, identifying eradication, containment and coexistence as modes of ordering grounded in distinct ethical assumptions. They show that uncertainty arises not from ignorance but from knowing differently and that coexistence requires navigating tensions between control and connection.

These dynamics shape everyday farm life. Farmers' routines of feeding, observing and treating animals are both affective and instrumental (Fry, 2023). Interspecies belonging captures how farmers and animals co-constitute places and identities through shared rhythms of labour and care. This familiarity forms what Schneider and Bubeck (2025) call *relational expertise*; tacit knowledge arising from co-presence rather than abstract metrics. Recognising this expertise allows governance frameworks to become adaptive and responsive. Production diseases can then be read not as discrete pathologies but as indicators of imbalance within relational ecologies (Buller et al., 2018; Hinchliffe et al., 2016). Health is defined not by the absence of disease but by the presence of sustaining relations within specific ecological and economic contexts.

Re-centring agricultural and veterinary practice in this way shifts attention from treating symptoms to cultivating systems capable of mutual flourishing, where care becomes the dominant organising principle. The persistence of production diseases does not simply signal technical failure; it reveals the limits of human-centred rationality and calls for a rethinking of what it means to live and die well together in agri-food systems shaped by business and management operations underpinned by CSR strategies (Amos & Sullivan, 2018; Janssens, 2021; Lever & Evans, 2017).

Expert collaborator and case

This chapter draws on the research of Lewis Holloway, a geographer at the University of Hull, whose work bridges animal studies, rural sociology and science and technology studies. As part of a Wellcome Trust project on bovine viral diarrhoea (BVD) in cattle and lameness in cattle and sheep, in the UK, Lewis and colleagues collaborated with veterinarians, epidemiologists and historians to examine how livestock health is co-produced through everyday relations and institutional discourses across farms, advisory networks and regulatory frameworks. As he notes in our interview, social science contributes to these debates by examining "how farmers acquire, adapt and apply knowledge about livestock diseases, often in ways that contrast with scientifically recommended best practices." Interdisciplinarity, he adds, reveals the "complex web of historical, social, economic and scientific factors that influence livestock health management."

Scholarship produced by Lewis and his collaborators (Holloway et al., 2023a, 2023b, 2024a, 2024b; Mahon et al., 2024) exemplifies ethical and epistemological reflexivity, foregrounding the ambivalent realities of farming amid market pressures, audit regimes and policy incentives. Their work shows how farmers, veterinarians and animals negotiate health within competing demands, revealing livestock health as a relational accomplishment rather than a technical output. Incorporating these insights allows this chapter to move beyond managerial framings towards a grounded account of health as relationally enacted through multispecies entanglement within contemporary systems of business governance. As Tallberg et al. (2024) argue, "multispecies inclusivity" offers significant potential for addressing entrenched problematic practices.

Conceptual foundations: Corporate biopolitics and animal health

Animal health and welfare scholarship has long been structured by rationalities that place food productivity and security at the centre of livestock concerns, and treat animals as units of biological and economic optimisation. From the 18th century onwards, veterinary medicine developed as an instrument of statecraft and economic management (Yeates, 2013). In Foucault's (2007) terms, this represents a biopolitical project focused on the regulation of life to optimise populations for human prosperity. Hinchliffe et al. (2016) usefully extend Foucault's analyses of power into the more-than-human domain by examining how animals became governable bodies to be measured, modified and exploited (Holloway et al., 2009; Holloway & Morris, 2017). The modern veterinary regime, as Swabe (1999) notes, thus becomes a way of stabilising human societies by controlling animal life through an apparatus in which care and domination coincide and through which animal bodies are enrolled into economic and organisational systems of value creation (see Amos & Sullivan, 2018; Lever & Evans, 2017).

Corporate biopolitical logics remain embedded within business and management frameworks. One Health and One Welfare – interdisciplinary policy frameworks that seek to integrate human, animal and environmental health and wellbeing – are often presented as progressive responses to global risk and interdependence (Cox, 2022; Kaiser-Grolimund et al., 2025). However, One Health in particular can potentially function as a global infrastructure of surveillance and risk management, using language of "balance" to conceal the asymmetry through which animals and ecosystems serve as instruments for securing human life (Cox, 2022). Within diverse corporate contexts, health is conceived of as a measurable state rather than a relational process, while welfare is reduced to the prevention of disease transmission to people and to the management of risk within supply chains (Amos & Sullivan, 2018). In a similar vein to Hinchliffe et al. (2016) and Swabe (1999), Cox (2022) calls for a new development paradigm focused on "flourishing" to replace the current "extractive" model within which living systems are valued for productivity and data yields and through which veterinary knowledge sustains capitalist accumulation by translating moral concern into technical rationalities operationalised through audits, metrics and assurance schemes.

For Kaiser-Grolimund et al. (2025), One Health's foundational problem is the assumption that biomedical science provides a universal account of life and health. Drawing on anthropology's ontological turn, they argue that communities differ not only in how they represent the world but in what worlds they inhabit. From this perspective, health is enacted differently within human and nonhuman relations, where animals may be regarded alternatively as persons or as kin. One Health's universalism marginalises these realities, they argue, reducing local knowledge to culture or belief. Kaiser-Grolimund et al. (2025) thus advocate for ontological openness, what we might think of as the capacity to engage multiple worlds of health without subsuming them under dominant bodies of knowledge or subordinating them to managerial abstractions.

These brief insights reveal deep-seated material and knowledge-based foundations to anthropocentrism in animal-health scholarship that extend into contemporary business, policy and governance practices (Köllen & Schneeberger, 2023; Twine, 2024). Materially, biopolitical regimes convert care into control, a process through which animal welfare becomes a technology of normalisation that aligns life with economic imperatives, silencing plural understandings of interspecies coexistence. Both forms render animals visible as data but invisible as subjects. The treatment of production diseases illustrates these dynamics clearly. Within industrial agriculture, such pathologies are understood as technical failures that require correction rather than as systemic contradictions within production-oriented organisational models. While anthropocentric scholarship demonstrates how life has become governable (Hinchliffe et al., 2016; Swabe, 1999), it rarely imagines how the health of livestock may be conceived not as an instrument of human survival or business continuity, but as a shared condition of "multispecies inclusivity" with direct implications for responsible management and ethical governance (Tallberg et al., 2024).

Towards an interspecies approach to livestock health and welfare

To include animals in the analysis of livestock health and welfare is not simply to acknowledge their presence. It is to recognise them as participants, actors whose bodily capacities, affective responses and resistances co-produce health outcomes within economic systems sustained by animal labour (Porcher, 2017). The challenge here, then, is not simply about adding animals in. It is to treat "species" as a sociological category on a par with class or gender in its own right (Wadham et al., 2024) and to recognise its relevance for management, governance and responsible business practice. Much social science positions animals as background to human labour or culture, while an interspecies approach foregrounds how humans and animals make farming worlds collaboratively as interdependent organisational collectives.

Including livestock as farming participants

The social science research undertaken by Lewis and his team embodies this shift, with the emphasis on placing human–animal relations at the centre of inquiry in ways that challenge top-down managerial assumptions about the business of farm animal welfare (Amos & Sullivan, 2018). This demands attention to the embodied, sensory and ethical dimensions of disease management; for example, how farmers see, feel and respond to their animals and how animals, in turn, communicate through movement, posture and behaviour within everyday work routines.

Hinchliffe et al.'s (2016) concept of "disease situations" offers useful insights here. Rather than viewing diseases in a purely technical sense, they are described as relational effects that emerge from entanglements among species, infrastructures and environments. Lameness and BVD are not simply biological events; they are configurations of bodies and management practices. An interspecies approach begins by reframing disease from an outcome of human mismanagement to a relationally enacted condition shaped by both human and nonhuman agency within production systems.

Lewis and colleagues explore these entanglements in northern English farms, analysing how farmers, veterinarians and animals collectively practice care around endemic conditions like BVD and lameness (Holloway et al., 2023a). Drawing on insights from around 50 interviews with farmers and advisors (including vets, specialist foot trimmers and so on), he and his collaborators show that care is an embodied practice rather than an abstract ethic or formalised policy ideal. As an interviewee stated when discussing a cage (a "crush") that is used to hold cows still for examinations and veterinary treatments: "The cow that is in the crush, that's all that should matter at that moment. There is nothing else that matters, only that cow... handling that cow with care and respect... the cow comes first" (Adviser 2, quoted in Holloway et al. (2023a, p. 1284). This moment of tactile, situated care is central to what Haraway (2008) calls *response-ability*, the capacity to notice and respond to another's needs in real-time organisational practice. At the same time, Lewis' team caution against romanticising such intimacy. Care in farming, they note, coexists with harm, making it more likely that endemic conditions persist (Holloway et al., 2023a). Acts of care (hoof trimming, culling, vaccinating) can simultaneously alleviate suffering and reinforce the systems that produce it within economically constrained production models.

The concept of "killability" (Giraud, 2019; Haraway, 2008) captures this paradox well. Although resisted by some farmers, Holloway et al. (2023a, p. 6) note that "a key part of lameness-reduction strategies is that persistently lame animals are culled from herds/flocks." Protecting the rest of the herd is justified as responsible management, yet this reveals the biopolitical logic through which life is governed at the level of populations and rendered compatible with efficiency and productivity goals (Hinchliffe et al., 2016). Including animals in the analysis means grappling with this moral ambivalence and recognising how welfare and violence, empathy and efficiency, are relationally enacted in everyday farming practice and business decision-making.

Animal agency and the co-production of disease

An interspecies perspective reveals how animals' own capacities and behaviours shape health outcomes and influence organisational responses. For instance, Holloway et al. (2023b, p. 189) document how cows and sheep resist or comply with interventions. Lameness, for example, is not simply detected but performed through the animal's gait and posture. Lewis' team refer to these assemblages as heterogeneous biosocial collectivities (Holloway & Morris, 2012; Morris & Holloway, 2014), which suggest temporary alignments of farmers, vets and animals coalescing around a shared biological concern within situated management contexts.

Within these collectivities, agency is distributed rather than owned. For farmers, recognising lameness involves observing how pain alters an animal's behaviour and responding accordingly. One farmer illustrated this by recounting a particular occasion when he identified a lame animal through behavioural changes:

> [When] we went to feed the animals, all the rest came forward to the trough and [one bullock] was standing at the back of the yard holding its foot up. Clearly, it'd lost its appetite or lost its ability to move around.
>
> *(Holloway et al., 2024b, p. 188)*

A cow's movement, as this example illustrates, can signal pain, so including animals means recognising how disease is not merely managed by humans but enacted through interspecies cooperation and constraint within organisational routines. Communication is key here, as a farmer noted: "You can tell with an animal's face when they're ill… if you know your animals" (Holloway et al., 2023a, p. 1285).

The team developed this argument further by linking interspecies health to commodification within capitalist economic systems (Holloway et al., 2024b). Drawing on Appadurai's (1986) concept of "commodity situations," they show how farmed animals occupy unstable positions between life and value. Their "liveliness," that is, their ability to grow, reproduce and produce milk, creates economic worth, but also exposes them to injury and disease within value-extracting systems. Following Shukin's (2009) analysis of "animal capital," animal health becomes a site where life is rendered economically productive through managerial and market logics. Lewis and colleagues (Holloway et al., 2024b) argue that endemic disease confirms that animals are valued for productivity, not wellbeing, and that animal "disability" (Taylor, 2017) is managed through culling or breeding rather than care, all of which has consequences for the welfare of individual animals and the broader management practices and economic viability of farms (Holloway et al., 2024b). Recognising animals as lively but vulnerable commodities invites a rethinking of value itself, one grounded in relational wellbeing rather than exchange value, with embodied familiarity forming part of what the team refer to as intersubjective and relational performance (Holloway et al., 2024a).

What inclusion changes

Including animals explicitly in our analysis changes both the epistemology and ethics of livestock health. Epistemologically, it challenges the dominance of management metrics and foregrounds experiential, sensory and place-based knowledge as legitimate forms of organisational knowledge. Farmers' observations such as "You can tell if they're not bright in the eye," or "Footrot is stinky and horrible," become legitimate data about health. Such local expertise complicates hierarchies of knowledge that privilege scientific over practical knowing. Ethically, inclusion shifts the focus from control to coexistence (Holloway et al., 2023a). As Wadham et al. (2024, p. 171) emphasise, rural life is "organically embedded in an interspecies world." Recognising this embeddedness reframes welfare not as a human gift but as a shared condition. Interspecies approaches therefore entail cultivating what Tronto (2020) calls political care, acknowledging how privilege and vulnerability are distributed across species, labour and organisational roles (Coulter, 2016).

At the policy level, Lewis notes that initiatives such as the UK Department for Environment, Food and Rural Affairs' (Defra) Animal Health and Welfare Pathway for England mark a move towards preventative, partnership-based approaches (see Defra, 2025). During our interview, he recognised that this "reflects a growing recognition that endemic diseases are not just animal welfare concerns but also economic and environmental issues that reduce cattle productivity." However, when veterinarians visit farms to assess cattle health and recommend interventions, the emphasis, he explained, remains on compliance and measurable improvement rather than lived wellbeing. An interspecies framework, in contrast, would assess success by the quality of relationships built around trust, attentiveness and adaptability between humans and animals rather than performance indicators alone.

Towards interspecies governance futures

What would governance look like if animals were treated as co-producers of health? Building on insights from Hinchliffe et al. (2016), governance would move beyond enforcing compliance towards facilitating responsiveness, recognising that health emerges not from regulation alone but through the alignment of human intentions with animal behaviours, microbial ecologies and material infrastructures. In this view, interventions are not imposed from above but negotiated through situated relations of care, observation and adaptation within interspecies organisations. In practice, this means that health management must attend to how farmers, veterinarians and animals co-constitute knowledge. Strategies for managing lameness, for example, could integrate farmers' sensory expertise and animals' behavioural signals alongside digital data and veterinary metrics (Holloway et al., 2024a). Such approaches embody what Harbers and Pols (2015) and Mol et al. (2010) describe as *"tinkering"*; that is, ongoing, improvisational adjustments sustaining multispecies wellbeing within everyday constraints.

Interspecies governance also resists the tendency to generalise from population-level data to individual lives, or to extrapolate individual welfare to systemic wellbeing. Attending to the histories, movements and embodied responses of particular cows and sheep prevents care from collapsing into compliance. It foregrounds a form of policy that begins not from abstraction but from lived encounters. As the empirical work of Lewis and his team demonstrates, the challenge is to build multispecies futures that balance privilege and vulnerability across species and systems. Including animals in governance is not simply an act of empathy; it is an act of epistemic justice that allows their experiences to shape what counts as knowledge, responsibility and welfare in organisational decision-making.

This shift entails reimagining institutions and organisations around co-production and "multispecies inclusivity" (Tallberg et al., 2024). Instead of treating farmers as implementers of policy and animals as data points, a responsive form of governance would view all participants as contributors to the collective health of the system. It invites participatory forms of decision-making, where farmers' experiential knowledge and animals' behavioural feedback inform welfare and disease-control strategies. Policy mechanisms such as Defra's Animal Health and Welfare Pathway could evolve to include reflective, qualitative accounts of care, complementing quantitative performance indicators. Lewis and colleagues' work carries profound implications for business and management in agri-food systems that depend on the health and welfare of animals. Their analysis reframes farms not as sites of linear control but as assemblages of interspecies management, where animal bodies, behaviours and capacities directly influence productivity and organisational outcomes. Recognising livestock as co-constitutors of value challenges conventional business metrics based solely on efficiency or yield, extending notions of performance to include responsiveness, adaptability and relational integrity.

This reframing invites a more dynamic and ethical conception of management. Managing endemic conditions such as lameness or BVD requires flexibility rather than standardisation, what management theorists might call dynamic capabilities (Pitelis & Wang, 2019), across species boundaries. The concept of interspecies care used by Lewis' team resonates with Tronto's (2020) idea of political care, with responsibility distributed unequally yet co-constituted through relational labour. Such care entails recognising that good management is not merely technical but ethical, requiring attentiveness to the more-than-human conditions sustaining production. For larger producers, processors and retailers, these insights point towards the value of relational governance models that embed animal-centred care, moving beyond compliance towards accountability and learning.

As Lewis notes during our interview, at a time when farming is increasingly shaped by business and management principles, this body of research extends the scope of business ethics into interspecies accountability. Responsible management must coordinate care, value and vulnerability across networks of human and nonhuman actors. Innovation, then, is not merely technological but relational: It lies in building infrastructures and cultures that enable attentiveness, negotiation and shared flourishing. Ultimately, an interspecies approach to the governance and management of farmed animals redefines what it means to live and work well together in systems of necessary dependence and inevitable harm. It does not deny the realities of production, death or commodification, but situates them within relations of care and accountability. As Giraud (2019) and Haraway (2016) observe, caring with animals entails remaining with the trouble of shared vulnerability, acknowledging that welfare is not the absence of harm but the presence of response.

Donaldson and Kymlicka's (2011) political theory of animal membership suggests that dependency generates governance obligations and responses. The scholarship of Lewis and colleagues demonstrates that including animals in this way reframes health as co-constituted, welfare as a relational achievement and management as shared *response-ability*. Such a shift will not eliminate the contradictions of livestock farming, but it will make them visible, discussable and governable, transforming endemic disease from a technical problem into an invitation to cultivate more liveable, attentive and just worlds characterised by forms of multispecies inclusivity that move beyond dominant anthropocentric norms (Köllen & Schneeberger, 2023; Tallberg et al., 2024; Twine, 2024). In this sense, livestock welfare is not simply a sustainability issue but a question of responsible management, revealing how business systems organise life, vulnerability and value across species.

Directions for future research

Future research on livestock health and welfare should advance a deeper understanding of how interspecies relations actively shape both the emergence and management of disease within contemporary agri-food systems. As Holloway et al. (2024a) demonstrate, endemic conditions such as lameness and BVD are not isolated biological events but relational effects, co-produced through everyday encounters among farmers, veterinarians, infrastructures and animals. Attending to these disease situations (Hinchliffe et al., 2016) requires moving beyond behavioural or technical explanations to examine how health itself is enacted across multiple forms of human–animal collaboration, constraint and care embedded in organisational routines and governance arrangements.

One key direction is methodological. Future studies could expand the use of ethnographic, participatory and sensory methods capable of capturing the embodied and affective dynamics of livestock care. Combining ethological observation with in-depth ethnography could reveal how farmers and animals co-constitute health through gesture, rhythm and responsiveness (Fry, 2023; Schneider & Bubeck, 2025). Such work would illuminate how disease management practices reshape affective relations and moral reasoning on farms (Holloway et al., 2023a), while comparative research across species and production systems could show how welfare is not universal but situated within specific multispecies arrangements (Haraway, 2008; Puig de la Bellacasa, 2017).

A second opportunity lies in connecting interspecies health to political economy. As Buller and Morris (2003) and Holloway et al. (2024b) suggest, endemic diseases are inseparable from the market logics that govern animal life. Research examining how farmers navigate overlapping ethical and economic pressures could help reorient welfare policy away from compliance towards what we might refer to as *relational accountability* and *responsive governance*. Interdisciplinary collaboration remains crucial here. Integrating veterinary science, social theory and animal behaviour research can generate more holistic frameworks for understanding how biological, affective and material forces interact in multispecies health management. Digital technologies such as precision monitoring should also be examined critically, not only as tools for optimisation but as mediators of attention and care. Following Tronto's (2020) call for political care, future research should ask whether such technologies enhance or erode farmers' capacity to respond to animals as sentient beings rather than as data points.

A final research direction concerns the temporal and emotional dimensions of care. Livestock farming involves enduring relationships marked by attachment, fatigue, loss and ethical discomfort (Giraud, 2019). Investigating how farmers live with empathy and ambivalence when caring for and culling animals would deepen understanding of interspecies ethics as lived organisational practice. As Haraway (2016) argues, living and dying well together requires acknowledging shared vulnerability rather than resolving it. Advancing interspecies research on livestock health therefore entails shifting attention from controlling disease towards cultivating responsiveness, reimagining health as an ongoing relational achievement sustained through care, adaptation and co-presence.

Ethical implications

Adopting an interspecies perspective on livestock health and welfare reconfigures ethics from a framework centred on human intention to one grounded in shared vulnerability and *response-ability* (Haraway, 2008). In the team's work, farmers' everyday encounters with cattle and sheep reveal care as a relational achievement enacted through touch, observation and responsiveness to animals' embodied signals rather than abstract moral commitments (Holloway et al., 2024a). Yet these same practices frequently involve harm: trimming hooves, culling chronically lame animals or restricting movement to prevent injury. Ethical life on farms therefore unfolds within what Giraud (2019) terms the ambivalence of care, where nurture and violence coexist.

From a feminist ethics of care perspective (Puig de la Bellacasa, 2017; Tronto, 2020), the challenge is not to purify care but to politicise it, to recognise how responsibility, vulnerability and power are distributed unevenly across species and systems. This aligns with Coulter's (2016) work on more-than-human labour, which highlights how animal bodies are enrolled into work relations that demand endurance while marginalising their vulnerability. An animal-centred ethic acknowledges that caring *for* livestock may involve actions that conflict with caring *about* them. Ethical reflection must therefore extend beyond individual intentions to the market logics, breeding regimes, veterinary protocols and policy frameworks that normalise disposability as responsible management.

In this view, ethics is less about avoiding harm than about sustaining care alongside unavoidable harm. It demands humility, reflexivity and openness to being affected by the lives and deaths of animals whose wellbeing remains entangled with human economic practice. Rather than functioning as a checklist of moral compliance in corporate infrastructures (Amos & Sullivan, 2018), interspecies ethics reframes welfare as an ongoing negotiation of coexistence within shared, unequal and fragile multispecies worlds.

Practical implications for responsible management

Engaging with livestock farming as a site of reform does not imply endorsement of animal agriculture as ethically unproblematic. Abolitionist and vegan scholars argue that systems premised on breeding, confinement and killing cannot be rendered just through managerial improvement alone. This chapter does not seek to resolve that debate. Rather, it examines how responsibility is negotiated within existing agri-food systems, while recognising that for some, more fundamental transformations – including transitions away from animal-based production altogether – remain ethically imperative.

For business and management, an interspecies perspective challenges what it means to manage responsibly in food systems. Integrating an interspecies perspective into livestock health and welfare entails a practical shift from treating disease as an isolated biological malfunction to understanding it as a relational imbalance within more-than-human organisational systems. Rather than viewing animals as passive biological resources (Amos & Sullivan, 2018; Hinchliffe et al., 2016; Shukin, 2009), this approach recognises them as participants in organisational and economic life whose bodily capacities and behaviours co-produce health outcomes. In terms of relational management, this means care, accountability and governance must be considered across human–nonhuman boundaries. As we can see in the study of cattle and sheep carried out by Lewis and his colleagues, farmers' capacity to notice subtle changes in behaviour (e.g. altered gait, reluctance to feed, withdrawal) emerges as central to effective care. Responsible management therefore begins by cultivating attentiveness and the capacity to interpret animals' embodied cues as forms of communication rather than mere technical data.

This reframing requires rethinking how welfare is monitored and governed. Contemporary corporate welfare regimes rely heavily on quantifiable indicators (Amos & Sullivan, 2018). While valuable, these metrics risk detaching care from lived encounters, transforming ethical responsiveness into procedural compliance. An interspecies approach challenges such productivity- and compliance-driven models by foregrounding health as relationally enacted through everyday human–animal interactions shaped by market pressures, policy instruments and ethical ambivalence. A relational model of management would therefore complement technical measures with qualitative attentiveness grounded in farmers' experiential knowledge, veterinarians' sensory expertise and animals' behavioural feedback. As Schneider and Bubeck (2025) argue, relational care depends on ongoing adjustment or *tinkering* (Harbers & Pols, 2015), small, situated improvisations that sustain wellbeing within everyday constraints.

Implementing this approach at scale requires institutional change. Advisory and policy frameworks such as Defra's Animal Health and Welfare Pathway could, in principle, incorporate participatory audits that include farmers' observations and reflections on animal behaviour, alongside established benchmarks. Such arrangements would support learning-oriented governance rather than punitive assessment, shifting responsibility beyond audit towards relational accountability across species. An interspecies perspective also reshapes the deployment of digital technologies. Precision livestock tools can either distance farmers from animals or enhance communication, depending on how they are designed and used. Sensors and monitoring systems should extend rather than replace embodied observation, alerting farmers to subtle deviations that prompt closer engagement. As Haraway (2016) and Tronto (2020) remind us, good care depends on remaining present and accountable within complex relations rather than outsourcing responsibility to technological fixes.

Finally, responsible management must address the emotional and ethical labour of farming. Caring for animals while accepting their eventual death requires emotional literacy and institutional support. Recognising the ambivalence of care (Giraud, 2019) as integral rather than exceptional can foster more resilient farming cultures. Training, peer dialogue and reflective practice can help farmers and veterinarians navigate interspecies responsibility without retreating into detachment or guilt. A sustainable welfare system thus depends not only on technical control but on cultivating trust, responsiveness and accountability across species.

Reflective activities inside/outside the classroom

These reflective activities are designed to help students, colleagues and the wider public think critically about what it means to treat animals as participants in the co-production of health and welfare. They aim to move learners from conceptual understanding to lived reflection, encouraging embodied awareness, ethical dialogue and contextual sensitivity beyond academic settings.

A Revision questions that test knowledge ("what?")

- A1 What is meant by a "veterinary regime"?
- A2 What are production diseases and why are they described as endemic?
- A3 How does corporate social responsibility (CSR) frame animal welfare within agri-food supply chains?
- A4 What role does farmers' experiential knowledge play in livestock health management, according to Holloway and colleagues?

B Reflective questions that build understanding ("so what?")

- B1 What ethical and political questions arise when examining power relations within livestock farming systems?
- B2 How does *tinkering* differ from technical, compliance-driven problem-solving?
- B3 In what ways do economic and policy structures influence who or what becomes killable in farming systems?
- B4 If welfare is co-produced through interspecies relations, what does this imply for how responsibility should be distributed across farmers, corporations and regulators?

C Take action ("do")

- C1 Spend time observing farmed or companion animals. Notice variations in posture, movement, breathing or social interaction. Reflect on what these signals might communicate and how your interpretations shift with sustained attention.
- C2 Make a diagram of routine decisions in livestock care (e.g. treatment, culling, feeding). Identify points of responsibility, vulnerability and agency for humans and animals.

- C3 Choose a small aspect of animal care such as cleaning, feeding or observing and adjust its timing, rhythm or method. Reflect on animal responses and what they reveal about interspecies negotiation and relational expertise.
- C4 Review a publicly available farm welfare policy (e.g. a retailer's animal welfare statement). Identify where it emphasises metrics and compliance and where it acknowledges relational care or animal agency.

D Ideas to discuss with colleagues, friends and family ("share")

- D1 How do scientific, regulatory or business frameworks shape what counts as "good welfare?"
- D2 Can multispecies perspectives coexist with commercial imperatives?
- D3 What forms of relational or experiential knowledge are undervalued in animal welfare management? How might these be recognised and legitimised within governance frameworks?
- D4 How does recognising animals as co-actors in health reshape everyday decisions about food, consumption or advocacy?

Key Ideas

Interspecies organisation

An interspecies organisation is a socially oriented production system in which humans and animals jointly shape routines, decisions and outcomes. Rather than treating animals as passive inputs, it recognises their agency, behaviour and embodied presence as organisational forces, meaning health, productivity and care emerge relationally through ongoing multispecies interaction and negotiation (Van Dam et al., 2023).

Responsive governance

Responsive governance replaces standardised welfare metrics with context-sensitive decision-making that values local knowledge, animal agency and situational flexibility. It recognises health as a dynamic relation shaped by diverse capabilities and constraints rather than a fixed biological state (e.g. Hinchliffe et al., 2016).

Relational accountability

As understood in this chapter, relational accountability extends responsibility beyond human institutions to the interdependencies with species that sustain production. It foregrounds attentiveness, trust and shared responsiveness among farmers, animals, advisors and regulators, emphasising that ethical management emerges from mutual engagement rather than hierarchical control (Haraway, 2016).

Political care

Political care situates everyday practices within broader structures of power. It highlights how economic pressures and policy frameworks shape who receives care and who bears harm, reframing care as an ongoing negotiation rather than a sentiment or technical requirement (Tronto, 2020).

References

Amos, N., & Sullivan, R. (Eds). (2018). *The business of farm animal welfare*. Routledge. doi:10.9774/gleaf.9781351270045.

Appadurai, A. (1986). *The social life of things: Commodities in cultural perspective*. Cambridge University Press. doi:10.1017/CBO9780511819582.

Bellet, C., Hamilton, L., & Rushton, J. (2021). Re-thinking public health: Towards a new scientific logic of routine animal health care in European industrial farming. *Humanities and Social Sciences Communications*, 8(1), 214. doi:10.1057/s41599-021-00890-y.

Buller, H., & Morris, C. (2003). Farm animal welfare: A new repertoire of nature-society relations or modernism re-embedded? *Sociologia Ruralis*, 43(3), 216–237. doi:10.1111/1467-9523.00242.

Buller, H., Blokhuis, H., Jensen, P., & Keeling, L. (2018). Towards farm animal welfare and sustainability. *Animals*, 8(6), 81. doi:10.3390/ani8060081.

Buller, H., Blokhuis, H., Lokhorst, K., Silberberg, M., & Veissier, I. (2020). Animal welfare management in a digital world. *Animals*, 10(10), 1779. doi:10.3390/ani10101779.

Coulter, K. (2016). *Animals, work and the promise of interspecies solidarity*. Palgrave Macmillan. doi:10.1057/9781137558800.

Cox, J. (2022). *Operationalising One Health–One Welfare*. www.wellbeingintlstudiesrepository.org/hw_onehealth/5.

Defra (UK Department for Environment Food and Rural Affairs). (2025). *Policy paper: Animal health and welfare pathway*. www.gov.uk/government/publications/animal-health-and-welfare-pathway/animal-health-and-welfare-pathway.

Donaldson, S., & Kymlicka, W. (2011). *Zoopolis: A political theory of animal rights*. Oxford University Press.

Enticott, G., Donaldson, A., Lowe, P., Power, M., Proctor, A., & Wilkinson, K. (2011). The changing role of veterinary expertise in the food chain. *Philosophical Transactions of the Royal Society B: Biological Sciences*, 366(1573), 1955–1965. doi:10.1098/rstb.2010.0408.

Foucault, M. (2007). *Security, territory, population: Lectures at the Collège de France 1977–1978*. Palgrave Macmillan.

Fry, T. (2023). "They're part of what we are:" Interspecies belonging, animal life and farming practice on the Isle of Skye. *Environment and Planning E: Nature and Space*, 6(4), 2495–2515. doi:10.1177/25148486231151809.

Giraud, E. H. (2019). *What comes after entanglement? Activism, anthropocentrism and an ethics of exclusion*. Duke University Press. doi:10.1215/9781478007159.
Haraway, D. J. (2008). *When species meet*. University of Minnesota Press.
Haraway, D. J. (2016). *Staying with the trouble: Making kin in the Chthulucene*. Duke University Press. doi:10.2307/j.ctv11cw25q.
Harbers, H., & Pols, J. (2015). Animal farm love stories: About care and economy. In A. Mol, I. Moser, & J. Pols (Eds), *Care in practice: On tinkering in clinics, homes and farms* (pp. 141–170). Transcript Verlag. doi:10.1515/transcript.9783839414477.141.
Hinchliffe, S., Bingham, N., Allen, J., & Carter, S. (2016). *Pathological lives: Disease, space and biopolitics*. Routledge. doi:10.1002/9781118997635.
Holloway, L., & Morris, C. (2012). Contesting genetic knowledge-practices in livestock breeding: Biopower, biosocial collectivities, and heterogeneous resistances. *Environment and Planning D: Society and Space*, 30(1), 60–77.
Holloway, L., & Morris, C. (2017). Biopower, heterogeneous biosocial collectivities and domestic livestock breeding. In M. Chrulew & D. Wadiwel (Eds), *Foucault and animals* (pp. 239–259). Brill. doi:10.1163/9789004332232_012.
Holloway, L., Morris, C., Gilna, B., & Gibbs, D. (2009). Biopower, genetics and livestock breeding: (Re)constituting animal populations and heterogeneous biosocial collectivities. *Transactions, Institute of British Geographers*, 34, 394–407 doi:10.1111/j.1475-5661.2009.00347.
Holloway, L., Mahon, N., Clark, B., & Proctor, A. (2023a). Living with cows, sheep and endemic disease in the north of England: Embodied care, biosocial collectivities and killability. *Environment and Planning E: Nature & Space*, 6(2), 1278–1298. doi:10.1177/25148486221105878.
Holloway, L., Mahon, N., Clark, B., & Proctor, A. (2023b). Changing interventions in farm animal health and welfare: A governmentality approach to the case of lameness. *Journal of Rural Studies*, 97, 95–104. doi:10.1016/j.jrurstud.2022.12.004.
Holloway, L., Mahon, N., Clark, B., & Proctor, A. (2024a) Interspecies encounters with endemic health conditions: Co-producing BVD and lameness with cows and sheep in the north of England. *Sociologia Ruralis*, 64(2), 180–201. doi:10.1111/soru.12458.
Holloway, L., Mahon, N., Clark, B., & Proctor, A. (2024b). Lively commodities and endemic diseases: Shifting commodity situations and nonhuman disability in cattle and sheep on UK farms. *Journal of Rural Studies*, 110, 103367. doi:10.1016/j.jrurstud.2024.103367.
Janssens, M. (2021). Animal business: An ethical exploration of corporate responsibility towards animals. *Food Ethics*, 7(1), 2. doi:10.1007/s41055-021-00094-9.
Kaiser-Grolimund, A., Bukachi, S. A., Karuga, J., Kämpfen, L., Keck, F., Zinsstag, J., & Brown, H. (2025). Does One Health need an ontological turn? *Critical Public Health*, 35(1), 2497358. doi:10.1080/09581596.2025.2497358.
Köllen, T., & Schneeberger, D. (2023). Avoiding unnecessary suffering: Towards a moral minimum standard for humans' responsibility for animal welfare. *Business Ethics, the Environment and Responsibility*, 32(4), 1139–1149. doi:10.1111/beer.12565.
Lever, J., & Evans, A. (2017). Corporate social responsibility and farm animal welfare: towards sustainable development in the food industry? In S. Idowu & S. Vertigans (Eds), *Stages of corporate social responsibility: From ideas to impacts* (pp. 205–222). Springer. doi:10.1007/978-3-319-43536-7_10.
Mahon, N., Finan, S., Holloway, L., Clark, B., & Proctor, A. (2024). Making a mark on the farm: The marks and traces of farm animals and infectious diseases in northern England. *Scottish Geographical Journal*, 140, 508–532. doi:10.1080/14702541.2024.2343951.

Mol, A., Moser, I., & Pols, J. (Eds). (2010). *Care in practice: On tinkering in clinics, homes and farms*. Transcript Verlag. doi:10.14361/transcript.9783839414477.

Morris, C., & Holloway, L. (2014). Genetics and livestock breeding in the UK: Co-constructing technologies and heterogeneous biosocial collectivities. *Journal of Rural Studies*, 33, 150–160. doi:10.1016/j.jrurstud.2012.10.003.

Norwood, F. B., & Lusk, J. L. (2011). *Compassion, by the pound: The economics of farm animal welfare*. Oxford University Press. doi:10.1093/acprof:osobl/9780199551163.001.0001.

Pitelis, C., & Wang, C. L. (2019). Dynamic capabilities: What are they and what are they for? *British Journal of Management Virtual Special Issue: The Landscape of Dynamic Capabilities*, 30, E1–E6. doi:10.1111/1467-8551.12346.

Porcher, J. (2017). *The ethics of animal labour: A collaborative utopia*. Springer. doi:10.1007/978-3-319-49070-0.

Puig de la Bellacasa, M. (2017). *Matters of Care: Speculative ethics in more-than-human worlds*. University of Minnesota Press. doi:10.1017/S2753906700002096.

Schneider, A. K., & Bubeck, M. J. (2025). The dual character of animal-centred care: Relational approaches in veterinary and animal sanctuary work. *Veterinary Sciences*, 12(8), 696. doi:10.3390/vetsci12080696.

Shortall, O., & Calo, A. (2021). Novel industry-government governance mechanisms for the eradication of bovine viral diarrhoea in the UK and Ireland. *Journal of Rural Studies*, 81, 324–335. doi:10.1016/j.jrurstud.2020.10.052.

Shukin, N. (2009). *Animal capital: Rendering life in biopolitical times*. University of Minnesota Press.

Swabe, J. (1999). *Animals, disease and human society: Human–animal relations and the rise of veterinary medicine*. Routledge. doi:10.4324/9780203265116.

Tallberg, L., Huopalainen, A., & García-Rosell, J. C. (2024). Beyond anthropocentrism: A call to action for multispecies inclusivity. *Business and Society*. doi:10.1177/00076503241271254.

Taylor, S. (2017). *Beasts of burden: Animal and disability liberation*. The New Press. doi:10.2307/jj.26193044.

Tronto, J. C. (2020). *Moral boundaries: A political argument for an ethic of care* (2nd ed.). Routledge. doi:10.4324/9781003070672.

Twine, R. (2024). *The climate crisis and other animals*. Sydney University Press. doi:10.30722/sup.9781743328996.

Van Dam, A., van Engelen, W., Müller-Mahn, D., Agha, S., Junglen, S., Borgemeister, C., & Bollig, M. (2023). Complexities of multispecies coexistence: Animal diseases and diverging modes of ordering at the wildlife–livestock interface in Southern Africa. *Environment and Planning E: Nature & Space*, 7(1), 353–374. doi:10.1177/25148486231160637.

Wadham, H., Schuurman, N., & Dashper, K. (2024). Editors' introduction to the special issue. Privilege, vulnerability and care: Interspecies dynamics in rural landscapes. *Sociologia Ruralis*. doi:10.1111/soru.12477.

Yeates, J. W. (2013). *Animal welfare in veterinary practice*. Wiley-Blackwell.

3

ANIMALS AND RELIGION

Introduction

Creation stories, cosmologies and religious rituals have long reflected and reinforced perceptions of animals across diverse cultural contexts, assigning symbolic or instrumental values that serve human spiritual, moral and socioeconomic needs (Durkheim, 1912/2001; Levi-Strauss, 1962; Robertson Smith, 1894). Despite the "posthuman" and "multispecies turn" in the humanities and social sciences in recent decades, studies of religion have remained largely anthropocentric (Halafoff et al., 2024a). This orientation draws from wider intellectual traditions that continue to centre human agency and moral authority, limiting the extent to which animals are recognised as ethical subjects rather than resources within religious worldviews.

Animal sacrifice provides a paradigmatic illustration of how such human-centric reasoning is enacted in religious practice. In ancient Judaism, sacrifice functioned as communal worship and divine appeasement (Douglas, 1962), while Christianity subsequently displaced material sacrifice with symbolic ritual forms such as the Eucharist. Further east, Buddhism and Jainism oppose animal sacrifice, whereas Hinduism maintains a more complex and internally varied relationship that cuts across ancient and modern human understandings (Dash et al., 2024; Fischer, 2023; Scanes & Chengzhong, 2018; Szűcs et al., 2012). In Judeo-Christian contexts more broadly, the concept of human "dominion over animals" (Genesis 1:26) has authorised a persistent subordination of animals to human needs (Kapek-Goodridge, 2024; Midgley, 1984), positioning them primarily as sacrificial rather than sentient bodies (DeMello, 2012). Although animal sentience has become a significant policy concern in Europe in recent decades, giving rise to notions of "sentient commodities" (Wilkie, 2005), theological revisionists advocating vegetarianism

DOI: 10.4324/9781003585541-5

within Christian (Linzey, 1987) and Islamic (Masri, 2009) ethics continue to situate animals within moral frameworks structured around human concerns for animal welfare (Szűcs et al., 2012).

In India, the relationship between ritual sacrifice in Hinduism and animal slaughter in an expanding meat industry exposes deep ethical, legal and political contradictions spanning the sacred and the profane. While industrial slaughter is legitimised through secular and economic rationalities, ritual sacrifice among lower-caste, rural communities is increasingly delegitimised by courts, reformist Hindu organisations and animal rights activists (Fischer, 2023; Govindrajan, 2018; Singh & Dave, 2015). These enduring human-centred framings continue to constrain the capacity of religious discourse to engage meaningfully with pressing ecological crises. These exclusions matter not only for theology and conservation, but for organisations across contemporary societies, where religious traditions continue to influence food systems, legal norms and public debates about animals and sustainability – even in ostensibly "secular" national contexts. Engaging with Salini Saha's ethnographic fieldwork on snakes in rural West Bengal in India, this chapter outlines a less human-centred approach that recognises animals as active participants in religious life.

Challenging dominant perspectives

During the late 20th century, critical animal studies scholars began to challenge the anthropocentric assumptions evident in religious discourse by both exposing and offering ethical responses to the "unprecedented subjection and exploitation of animals" (McCance, 2013, p. 4). This work laid the foundations for the "multispecies" or "posthuman" turn in the humanities and social sciences, a reorientation that foregrounds interdependence, nonhuman agency and the entangled lives of humans, animals, plants and ecosystems. Influenced by thinkers such as Donna Haraway (2003, 2007, 2016), Anna Tsing (2015) and Robin Wall Kimmerer (2013), this shift calls into question the very basis of human exceptionalism that underpins Western thought and dominant religious paradigms. Instead of seeing animals as passive agents of human exploitation, posthumanist and multispecies approaches emphasise relationality, affect and co-becoming.

These frameworks also intersect with indigenous epistemologies, where animals are seen as kin, communicators and divine manifestations and not simply as metaphors or indicators of morality (Weatherdon, 2022). This allows us to ask not only how humans represent or think about animals, but how animals themselves participate in the creation and experience of the sacred. It also invites us to see animals not just as objects of doctrine, but as subjects of spiritual life (Halafoff et al., 2024b). This is not simply a theoretical move; it also has methodological implications. Methods such as multispecies ethnography and participatory, place-based research enable scholars to trace how animals shape religious practice, identity and ecological ethics on the ground (Halafoff et al., 2024a, 2024b). In this chapter, we adopt such a framework to examine religion as something fundamentally multispecies and posthuman, emphasising a dynamic, lived process that includes animals not only as symbols, but as agents and collaborators. This allows us to challenge dominant human-centric narratives through more inclusive, ecologically attuned ways of knowing and being.

Animals and religious entanglements

A multispecies account of religion in India as articulated by Annu Jalais (2010) and Radhika Govindrajan (2018) challenges religious frameworks that have historically separated humans from animals morally and ontologically through binary distinctions. Such work does not follow a Western model of animal reverence; it emerges from a situated cosmology in which religious, ecological and social life are inseparably intertwined.

This context is particularly valuable for exploring the limitations of the human–animal binaries that structure much religious thinking in the West and beyond. The practices involved, as we shall see, are not metaphorical in a religious sense; they are grounded in embodied relationships and local knowledge. This strand of work illustrates how religion is lived not simply through scripture or institutional authority, but through multispecies cohabitation and mutual awareness. These insights challenge Western models of religion as primarily representational and open the door to a relational, multispecies religious ethics grounded in shared vulnerability and care.

Expert collaborator, case and context

Our expert collaborator in this chapter is Salini Saha, an early career researcher who conducted the work explored in this chapter as a PhD student in the Department of Sociology at Presidency University, Kolkata, India. Her interdisciplinary background (spanning sociology, anthropology, religious studies, political economy and conservation studies) is highly innovative and allows her to provide far-reaching insights. Salini's doctoral research in West Bengal was based on a deeply contextualised, multispecies ethnography that attends to the lived experiences of humans and snakes in a shared agrarian environment. Salini's research highlights animals' multiple economic, ritual and symbolic roles through their entanglement in debates about conservation, ecology and tourism. In a study of village deities and agrarian landscapes, for example, she extends this analysis beyond snakes to show how everyday labour, caste relations and land-based livelihoods shape religious practice (Saha, 2025). This work highlights how sacred life is embedded in contested political and economic structures rather than confined to theological doctrine.

Building on insights derived from Global South feminist and postcolonial scholars (Govindrajan, 2018; Jalais, 2010), Salini's (Saha, 2024, 2026) work exploring snakes (i.e. cobras, known locally as *Jhanglai*) in rural West Bengal offers a compelling case through which to explore the ideas outlined above. In the Bardhaman district where she conducted her fieldwork, these snakes are not simply treated as wild animals to be feared or eradicated. They are understood as sacred beings with whom villagers share everyday spaces and cultivate ongoing social relationships.

Salini's methodology reflects a posthumanist and decolonial ethos that refuses rigid disciplinary frameworks, weaving together ethnographic observation, materialist analysis and local cosmologies. By centring the cobras not as symbols or resources, but as co-actors in social and spiritual life with low-caste Hindu groups in rural areas, Salini challenges the extractive logic that continues to dominate religious and ecological studies. Her adoption of the notion of *"otherwild spaces"* (Govindrajan, 2018) that are neither "wild" nor "domesticated" (as they are in colonial and conservation discourses) offers a powerful framework for thinking about how "multispecies religion" is shaped by mutual, ongoing entanglement. It resonates with broader calls to recognise indigenous and subaltern epistemologies that refuse to separate the sacred from the ecological, a form of "subaltern environmentalism" that refuses to separate the spaces of daily life from nature (Armiero & Sedrez, 2014).

While Hinduism is often celebrated for its reverence for animals, especially the sacred cow (Harris, 1966), it still accommodates sacrificial practices. India's largest animal rights group, *People for Animals*, argue that sacrificial practices among lower caste, rural groups were influenced by Islamic and Christian theology, which points to the complex and politically contested position of animals in India that cuts across sacred and profane forms (Singh & Dave, 2015). Despite the national myth that India is a vegetarian nation, Fischer (2023) shows that meat production and consumption are expanding in India even in the face of climate change and wider sustainability concerns. These developments are problematic, not least because they obscure the complex entanglements between humans and animals in spiritual, social and environmental life, impeding the development of ethical practices that acknowledge animals as co-inhabitants of the earth (Govindrajan, 2018; Saha, 2024, 2026). Understanding these interspecies religious dynamics, then, is essential for organisations engaged in tourism, conservation, rural development and climate adaptation, where sacred animals are not peripheral but central to ethical and economic life.

In what follows, we outline the anthropocentric foundations of animals in studies of religion, before looking at how these conceptual foundations are being challenged. Turning to the work of Salini Saha and others, the chapter then explores what it means to include animals in a lived, grounded account of religion. Finally, we discuss future research possibilities that embrace animals in multispecies accounts of religion and how this might enable more inclusive and responsible management practices.

Prioritising human superiority through religion

Much scholarship on animals and religion is grounded in conceptual frameworks that prioritise human superiority through creation stories. This worldview is rooted in both Western philosophical traditions and in dominant theological interpretations, particularly within the Abrahamic faiths. One of the most influential sources is the Biblical notion of "dominion" from Genesis 1:26–28. Historically, this has been interpreted as divine sanction for humans to use animals for food, labour and religious purposes. While more recent interpretations move towards a stewardship model emphasising care and responsibility, the dominant position remains focused on human interests and needs (Caruana, 2020; Pope Francis, 2015).

Philosophically, anthropocentrism was reinforced through the thinking of Aristotle, who ranked humans above animals because of their capacity for rational thought and used them in experiments for the human benefit (Dunn, 2006). This hierarchical structure was further elaborated by Thomas Aquinas (1225–1274), who argued that because animals lack immortal "souls" they are morally inferior. Although Aquinas condemned cruelty to animals, his reasoning was that cruel acts might cultivate moral insensitivity in humans rather than harm to animals themselves (Szűcs et al., 2012; Wade, 2004). This instrumental view, which sees animals as a means to human ends, has underpinned centuries of religious, ethical and legal thought.

In much religious scholarship, animals are thus seen in symbolic terms rather than as beings with moral worth. In Christianity, animals often appear in allegorical roles: lambs symbolise innocence, serpents represent evil. This pattern recurs across traditions, where animals are used to illustrate human virtues or vices, often erasing their lived realities (Caruana, 2020). Religious dietary laws such as *kashrut* in Judaism and *halal* in Islam include provisions to minimise animal suffering, yet these are generally framed around human spiritual purity and obedience to divine law rather than animal welfare *per se* (Freidenreich, 2011; Lever & Fischer, 2018; Masri, 2009; Szűcs et al., 2012). Even in revisionist theologies, views on animals remain largely anthropocentric (McLaughlin, 2012; Shavarani, 2024; Wade, 2004).

This is also the case in contemporary sustainability discourses (Kopnina et al., 2025) where animals are protected primarily for the role they play in protecting ecosystems or enhancing human moral and spiritual life. As Pope Francis (2015) stated in *Laudato Si'*, all creatures are moving towards God, although humans retain a unique moral status and responsibility. Anthropocentric hierarchies also persist in Eastern traditions such as Confucianism and Hinduism, although debate is often framed within the context of a cosmic order rather than dominion (Ip, 2025). While Hinduism venerates certain animals for their association with deities, it continues to place humans at the pinnacle of existence, notably through reformist food industry policies influenced by secular modernity and the politics of Hindu nationalism (Fischer, 2023; Govindrajan, 2018; Singh & Dave, 2015).

Contemporary scholarship is now critically examining the human-centred paradigm via relational and co-dependent models that recognise animals as agents and subjects in their own right. Mainstream religious thought often struggles to incorporate such perspectives, revealing the persistence of deeply rooted binary distinctions and conceptual assumptions. As the climate crises and a range of related concerns continue to highlight the consequences of these human-centred worldviews, the urgency of reassessing anthropocentric religious frameworks becomes increasingly necessary (Halafoff et al., 2024a, 2024b). Understanding these foundations is essential not only for academic critique, but for developing more inclusive and ecologically just religious ethics.

Religion and the interspecies turn

The interspecies turn in anthropological and sociological studies of religion invites us to rethink our ethical, cultural and epistemological relations with nonhuman animals. Moving beyond anthropocentrism, it urges scholars to recognise animals not merely as symbols or resources, but as agents and co-inhabitants of shared worlds. Salini's ethnographic work in West Bengal offers a powerful case study that exemplifies what it means to take animals seriously. Through her multispecies engagement with sacred serpents, particularly *Jhanglai*, Salini develops an in-depth account of interspecies dynamics, where animals are central rather than peripheral to religious life, ecological thought and local politics.

Beyond the symbolic and metaphorical

To include animals in the study of religion is to move beyond viewing them as metaphors, symbolic objects or passive recipients of human care. It is to acknowledge their agency, social presence and capacity to affect and be affected. Salini's work, rooted in multispecies ethnography and informed by posthumanist and decolonial insights (Govindrajan, 2018), frames animals as participants in religious cosmologies and everyday life.

In Musharu village that Salini visits during her fieldwork, the snakes are seen not as wild or domesticated, as they are in colonial and conservation discourses (Swanson et al., 2018), but as liminal beings that are sacred, feared, respected and intimately embedded in the village ecology. Villagers recount personal stories of cohabiting with *Jhanglai*, describing rituals that frame the snakebite as divine *prasad* (a sacred offering from a goddess) and even create architectural accommodations for these serpents within their homes. As one of Salini's interviewees explained when discussing an encounter with a cobra in her home:

> [I]t crept on to my back… I did not move. It will not bite if you don't move. If you move, it gets the feeling that you are going to harm it and it attacks. I waited until it left my body and then I saw it slowly slithering its way to the fields.
>
> *(Resident cited by Saha, 2024, p. 244)*

This relationship is not based on domination or taming but on negotiated coexistence, shaped by mutual vulnerability and shared dwelling. In these liminal zones and spaces, "where the snakes are described as *ghorowa* (homely) but not *posha* (pet)… there are no clear rules of domination and subordination between animals and humans" (Saha, 2024, p. 249).

To consider animals in this way is to see them as actors in religious and ecological fields, shaping human subjectivities, village ontologies and spiritual geographies. While conservationists and medical experts try to raise awareness about the danger of snakebites, the modern scientific treatments they offer evoke mixed feelings among villagers, where bathing in a local pond is seen as an appropriate treatment for a bite. Including animals also means incorporating their perspectives, even if those perspectives are inferred through human narratives, reactions and patterns of cohabitation. There is a considerable interpretive challenge accommodating the insights that emerge, but working in this way can lead to a rich relational account of religious life. In Musharu village, for example, residents believe that the snakes react to emotional and spiritual disturbances and that climate fluctuations disrupt not only the snakes' physical habitat, but their divine temperament too. As Salini explains in our interview, "they believe that the snakes are also the balancers… the harbingers of our climatic balance," and it is this spiritual ecology that informs both ritual and environmental responses.

Such encounters illustrate how interspecies religious life resists neat categorisation. *Jhanglai* is both a deity and a dangerous animal, a kin and a threat. This complexity reflects what Govindrajan (2018) calls the "*otherwild*," that is, a messy, relational wilderness that defies the binary logics of sacred/profane, for example, or wild/domestic found in anthropocentric accounts of religion and conservation (Durkheim, 1912/2001; Ingold, 2000). Salini uses this framing to argue that snakes are neither objectified as exotic specimens nor reduced to cultural symbols. They are vital participants in what she terms a "natural cultural world," where ecological survival and religious continuity are co-dependent.

This interspecies entanglement is particularly striking amid climate change. The declining monsoon and rising temperatures not only threaten agriculture; they also threaten the survival of the snakes. However, while conservationists frame the cobras as endangered species in need of extraction and protection, villagers see them as divine cohabitants whose disappearance would signify spiritual rupture as well as ecological loss. As one of Salini's (Saha, 2024, p. 348) respondents put it, "Hopefully they [the snakes] will [come out this year]… we are worried there may not be enough of them left in the coming years." This interspecies connection is deeply gendered and generational. Salini documents how newly married women, often recent arrivals, express both fears and fascination. Their attitudes highlight the process of cultural assimilation into a sacred ecology where snakes are both protectors and threats. This tension between affective unease and ritual respect is central to understanding how animal presence reshapes social and religious life.

Agency, place and knowledge

Salini's work challenges Western logics that separate humans from nature and animals from culture. Her interlocutors reject the idea of relocating snakes for scientific study, seeing such actions as violations of relational ethics. While scientists might emphasise classification and species management, villagers prioritise relational knowledge and spiritual guardianship. As Salini explains during our interview:

> There are the zoological surveys of India people, who believe that the snakes have to be taken away from human spaces in order to study and conserve them in a better way. And then there are the local inhabitants for whom… their religious life world is typically based upon the presence of the snakes.

These competing worldviews and frameworks reflect broader epistemological conflicts between indigenous ecological wisdom and state-sanctioned views on conservation and environmentalism about how best to understand and care for animal life (Govindrajan, 2018). In this way, Salini's study is not only about religion and animals; it is about power, legitimacy and contested futures.

Salini's (Saha, 2024) work also reveals the deep entanglement of agricultural practices such as farming and animal husbandry with spiritual life, particularly deity worship and sacrificial rituals. Animals in these contexts occupy multiple roles: Cows serve instrumental economic functions as plough animals and dung producers, while goats and cobras are caught in complex ethical and spiritual relations. Unlike cobras, however, which are deified and never sacrificed, goats are ritually killed during religious festivals to honour local deities. These seemingly contradictory practices coexist, as Govindrajan (2018) also argues, within a culturally coherent system of relational ethics grounded in reciprocity, obligation and affect, rather than in abstract notions of individual rights. The idea that goats may be sacrificed willingly, as kin or moral agents, underscores this affective logic and challenges liberal critiques of sacrifice as cruelty (Govindrajan, 2018). The tension between these embedded practices and state or reformist interventions was evident in a 2011 court ruling inspired by secular modernity, which prohibited religious sacrifice, reduced it to the realm of food production and erased its spiritual meaning (People for Animals v. State of Uttarakhand, 2011). Salini's work shows how these reformist interventions also reshape the moral and economic organisation of village life, disrupting relationships between caste, subsistence labour, land and sacred practice rather than simply regulating ritual behaviour (Saha, 2025).

Another original aspect of Salini's analysis of the entanglements between religion and economy is her focus on tourism and conservation. The rise of ecotourism, academic interest and media attention has transformed snakes into valuable economic resources. While outsiders express interest in developing tourism around the cobra deity *Jhanglai*, local villagers resist the overt commodification of sacred beings. As Salini noted during our interview, "they completely denied the claims of a tourism industry coming up in that place, despite clear interest from outside parties" (Saha, 2024). This mirrors Jalais' (2010, p. 211) account of the Sundarbans, where conservation narratives have framed tigers as "man-eaters" to satisfy tourist expectations.

By recognising cobras as agents in village economies, Salini argues for a shift in the tourist gaze, away from seeing animals as passive objects to viewing them as co-participants in social and economic life. Even where such beings generate economic value, their sacred status is maintained through relational ethics that resist full absorption into market logic. The cobras and goats discussed by Salini reveal that animal reverence and sacrifice are not oppositional, but part of a shared logic of interspecies interdependence shaped by agriculture, religion and rural economies. By engaging with figures like Govindrajan (2018) and Jalais (2010), Salini also situates her work within a Global South feminist and postcolonial framework, emphasising affective cohabitation and epistemological pluralism.

Towards an interspecies religious ethics

The religious worldviews of the villages Salini worked with are not static: They are shaped by and responsive to ecological pressures. As in Jalais' (2010) account of the Sundarbans, the villages of the Bardhaman district are conceived as a living and morally charged landscape populated by numerous beings with whom humans must constantly negotiate. The sacrality of snakes is not simply linked to a religious hierarchy, but to an adaptive, living cosmology that aligns with broader ecological ethics. An interspecies approach to animals and religion is thus not simply about adding animals into existing frameworks; it requires rethinking the frameworks themselves. It demands ethnographic attentiveness, theoretical openness and ethical reflexivity.

Through her immersive, relational and decolonial methodology, Salini offers a model for what such work can look like. The villages she conducts fieldwork in are not anomalies. They are sites of deep philosophical insight that reveal how spiritual life emerges not in isolation from the animal world, but through it. In highlighting the complex relationships between animals, climate, religion and identity, Salini's work encourages scholars to engage more seriously with nonhuman presence. It also calls for a pluralistic ethics that accommodates local cosmologies, ecological realities and lived religious practices. As environmental crises intensify, such grounded, interspecies perspectives are not only academically valuable; they are an urgent necessity.

What Salini's work reveals is not simply that animals matter in religious life, but that human religiosity is constituted through animal *presence*; snakes are not peripheral to religious devotion, they are central to it. This insight has profound implications for how we theorise religion, ethics and human–animal relations. It calls for an interspecies religious ethics that recognises mutual vulnerability, relational ontology and the inseparability of ecological and spiritual life. This raises broader methodological questions for religious and environmental studies. What kinds of methods allow us to take animal presence seriously, not just as backdrop but as constitutive of human religious life? Salini's fieldwork shows that deep immersion, unstructured interviews and attention to embodied practices (such as rituals, gestures and spatial arrangements) are key to capturing interspecies dynamics. Her work among villagers, attention to caste and gender dynamics and reflections on her own positionality as a researcher all strengthen her contribution to multispecies ethnography.

Directions for future research

Future research on animals and religion would benefit from further developing the interspecies framework introduced in this chapter, with a focus on comparative, transdisciplinary and decolonial methodologies. There is a pressing need to investigate religious practices not only in textual and doctrinal terms, but as multispecies, place-based experiences shaped by ecological conditions and local cosmologies. Ethnographic work that centres animals' roles in rituals, livelihoods and daily cohabitation across varied geographies in the manner of Govindrajan (2018), Jalais (2010) and Salini's own work (Saha, 2024, 2026) could generate new insights into religious multispecies ethics.

Further research could explore how global environmental pressures – particularly biodiversity loss and climate change – are reshaping religious understandings of human–animal relationships. Do new spiritual understandings emerge when ecological collapse threatens the presence of sacred species? How do these shifts vary across contexts of caste, class, gender and geography? Salini's work on climate anxiety in West Bengal provides a productive model for such inquiry. There is also scope for examining how state policies, conservation regimes and NGO interventions interact with local religious cosmologies, sometimes reinforcing, sometimes undermining traditional animal-related beliefs (Fischer, 2023; Govindrajan,

2018). Cross-cultural and cross-religious comparisons can clarify tensions and overlaps between institutional and community-level perspectives on animal welfare, conservation and ritual.

Comparative projects across the Global South and North would also reveal how religious narratives adapt to planetary crises. The veneration of sacred serpents in India, for example, might be compared with indigenous cosmologies in Latin America (Astor-Aguilera, 2016) or with Christian eco-theologies (Wyller et al., 2024) that frame animals as co-creatures in a shared creation. Such analyses would show how multispecies ethics operate across scales – from village shrines to global climate movements – and how they inform post-secular environmental politics. Developing new methods for studying animal agency in religious life is another important consideration. Traditional ethnography centres human voices, but multispecies and posthuman approaches invite more sensory and embodied methodologies.

There are also policy and applied research opportunities. As governments and NGOs increasingly frame biodiversity protection and climate adaptation in moral or spiritual terms, understanding how religious communities perceive animal agency and sacred ecology becomes crucial. Integrating local cosmologies into conservation and climate initiatives may yield more enduring outcomes than purely technocratic interventions. Likewise, exploring how interspecies ethics inform community resilience during ecological disruption could enrich adaptation and environmental justice frameworks. Gender and generation issues also merit closer attention. Women's ritual and care practices often mediate the most intimate human–animal encounters, as Salini (Saha, 2024) shows, while the intersections with climate justice and new forms of vegetable sacrifice also merit attention (Markowitz & Avieli, 2022). Comparative studies of these dynamics can reveal how spiritual and ecological knowledge circulates across generations and traditions.

Ethical implications

The ethical implications of implementing the interspecies approach outlined in this chapter are profound and multifaceted. At its core, this framework challenges entrenched anthropocentric moral systems that have historically viewed animals as symbolic, instrumental or subordinate to human needs and concerns. By contrast, the relational ethics proposed here emphasises animals as agents with spiritual, ecological and moral significance (Govindrajan, 2018; Jalais, 2010; Saha, 2024, 2026). This shift requires rethinking how ethical value is assigned, not through universalist notions of animal rights or animal welfare, but through context-specific, reciprocal relationships of cohabitation and mutual vulnerability.

Salini's work with sacred cobras in West Bengal illustrates the ethical depth of this approach. Villagers' relationships with snakes are not guided by abstract ethical codes, but emerge from everyday, embodied interactions grounded in reverence, fear, care and spiritual responsibility. Removing snakes from this context, whether for conservation or scientific study, does not simply constitute an ecological loss, it constitutes a moral rupture in the community's moral world.

Ethically adopting an interspecies perspective demands attentiveness to diverse places, groups and communities, an openness to plural moral frameworks and opposition to the assumption that one size fits all. It urges scholars, policymakers and practitioners to resist imposing dominant Western paradigms and instead engage with situated knowledge systems that honour the agency and presence of animals (Govindrajan, 2018; Jalais, 2010; Saha, 2024). This has implications for conservation ethics, religious reform, animal protection laws and cross-cultural dialogue. Ultimately, the approach advocates for an ethics grounded not in control or abstraction, but in shared entanglement, a vision of moral life in which humans and animals co-constitute each other's spiritual, economic and ecological futures.

Practical implications for responsible management

Introducing interspecies perspectives into responsible management challenges established business and organisational paradigms that position animals as resources, symbols or risks to be managed. This chapter suggests a reorientation in managerial thinking, one that considers animals as stakeholders (Tallberg et al., 2022) in socioeconomic and spiritual systems. In the contexts explored by our expert collaborator, sacred cobras are not mere objects of conservation or tourism; they are community members who shape decisions about land use, infrastructure and even healthcare. For managers operating in culturally and ecologically sensitive areas, this insight has far-reaching implications.

Organisations involved in ecotourism, conservation, or rural development must engage with local religious and ecological cosmologies. Attempts to commodify sacred animals for tourism, as seen in Salini's work, may be resisted or denied by communities who see such actions as sacrilegious. Responsible management in these contexts requires participatory decision-making that honours local beliefs and animal presence as active forces in shaping ethical and economic life.

Interspecies management also calls for new risk and crisis frameworks. In Salini's work, the presence or absence of snakes is seen as an indicator of ecological balance and spiritual wellbeing. Rather than treating animal populations as metrics of environmental health alone, managers should consider how communities interpret animal behaviour within moral and spiritual narratives. This approach can enhance culturally attuned climate adaptation, conservation and disaster response strategies. This perspective also has implications for workplace culture and leadership. A relational ethic that recognises human–animal entanglements can foster inclusive values and deeper environmental responsibility. This means that religious or ethical concerns about the treatment of animals in relation to halal or kosher certification (Lever & Fischer, 2018), or in regions with strong Buddhist and Hindu beliefs (Szűcs et al., 2012), must be recognised not just as compliance issues, but as moral commitments embedded in place-based worldviews.

Finally, this approach can encourage businesses to think beyond extractive or anthropocentric models of growth. Multispecies entanglements show that prosperity may depend not only on financial capital, but on ecological health and spiritual reciprocity (Govindrajan, 2018; Jalais, 2010; Saha, 2024, 2026). Responsible managers can learn from indigenous and subaltern models of economic life, in which animals, ancestors and deities are all part of the moral economy. These insights can inform sustainability strategies that centre mutual care, ecological embeddedness and ethical plurality. In sum, an interspecies perspective broadens the scope of responsible management by asking not just how to manage animals, but how to manage with them. It emphasises relational accountability, spiritual ecology and cultural specificity, essential components of responsible management in an era of overlapping ecological, economic and moral crises. In doing so, it positions interspecies religious ethics as central to thinking about sustainability and responsible management in culturally plural societies.

Reflective activities inside/outside the classroom

These activities are intended to deepen students' critical engagement with the interspecies approach and its ethical, spiritual and ecological implications. The aim is to foster not only conceptual understanding but also emotional and relational insight.

A Revision questions that test knowledge ("what?")

- A1 What does Salini Saha's case study of the *Jhanglai* cobras reveal about the lived entanglement of religion, ecology and social life?
- A2 How does Govindrajan define "otherwild" spaces?
- A3 How do these "otherwild" spaces differ from conventional ideas of wilderness?
- A4 How does Salini's research challenge conservationist approaches that frame snakes as species to be removed or managed?

B Reflective questions that build understanding ("so what?")

- B1 How does an interspecies religious ethic differ from anthropocentric ethical frameworks commonly found in religious traditions?
- B2 Why is the recognition of animal agency significant for religious and ethical thought?

- B3 In what ways can insights from an interspecies religious ethic inform sustainability strategies?
- B4 What might change in religious education or conservation policy if animal agency was to be taken seriously?

C Take action ("do")

- C1 Draw a diagram mapping your own daily entanglements with non-human life, whether as pets, food, clothing, urban wildlife, rituals or habits. Reflect on whether these relationships are exploitative, reciprocal, spiritual or indifferent. How might you reframe one of these relationships using a relational or interspecies ethic?
- C2 Visit a local religious or ecological site (a shrine, farm, urban park or temple) and observe how animals are present or absent. Note whether they are integrated into the space as sacred, symbolic, threatening or invisible. Write a short field note reflecting on how the space might be reimagined through an interspecies lens.
- C3 Keep a week-long journal documenting your interactions with non-human animals, whether in food choices, media, prayer, rituals or daily routine. Each entry should reflect on how these interactions are shaped by religious, cultural or ethical assumptions. Consider how a shift in perspective might change your actions.
- C4 Identify a recent public debate involving animals and religion (e.g. ritual slaughter, wildlife conservation, sacred species). Analyse how different actors frame the animal, i.e. as resource, symbol, threat, kin or stakeholder. What assumptions shape each position?

D Ideas to discuss with colleagues, friends and family ("share")

- D1 Ask people in your community how they think animals fit into their religious or moral worldview. Do they believe animals can sense divine presence, experience suffering spiritually or play a role in ritual beyond symbolism?
- D2 Discuss with colleagues or friends how scientific and indigenous/religious knowledge systems treat animals differently. What tensions arise when conservation or policy interventions ignore relational or sacred values attributed to animals?
- D3 With family or peers, debate whether animals should be considered stakeholders in ethical or environmental decisions. How might recognising them as moral agents alter views on farming, sacrifice, conservation or religious practice?
- D4 Talk with others about personal or cultural stories of cohabiting with animals. How do these stories shape your view of what counts as ethical or sacred?

Key ideas

Anthropocentrism

A worldview that positions humans as ontologically and morally superior within a hierarchical order of life. In religious contexts, anthropocentrism privileges human agency, salvation and meaning while marginalising nonhuman beings. This orientation shapes dominant theological narratives by obscuring multispecies entanglements and reinforcing human-centred interpretations of religious practice and ecological responsibility (Pope Francis, 2015).

Otherwild spaces

Relational spaces that unsettle the binary distinctions of wild/domestic and sacred/profane. Vis-à-vis sites of human domination or exclusionary conservation, *otherwild spaces* are constituted through negotiated coexistence, vulnerability and interspecies proximity. The concept critiques both colonial conservation regimes and orthodox religious frameworks by foregrounding situated, multispecies relations (Govindrajan, 2018).

Spiritual ecology

A framework in which ecological processes and religious meanings are mutually constitutive. Within spiritual ecology, sacred animals function as indicators of both environmental and spiritual wellbeing; their presence or decline carries ecological, emotional and theological significance. Animals are thus understood simultaneously as climate actors and divine mediators, collapsing the distinction between ecological change and religious transformation (Saha, 2024).

Ritual

Ritual refers to patterned, symbolic practices through which religious communities reproduce meaning, identity and moral order. In multispecies contexts, rituals involving animals can normalise sacrifice and hierarchy or cultivate care and relational accountability, depending on how they structure human–animal encounters (Jalais, 2010; Saha, 2024).

Sacred and profane

The sacred/profane distinction, associated with Durkheim, describes how societies separate what is revered from what is ordinary. Animals may be positioned as sacred beings, sacrificial offerings or profane commodities, shaping how their lives are valued, protected or exploited within religious and economic systems (Durkheim, 1912/2001; Govindrajan, 2018).

References

Armiero, M., & Sedrez, L. (Eds). (2014). *A history of environmentalism: Local struggles, global histories*. Bloomsbury.

Astor-Aguilera, M. (2016). Latin America: Indigenous cosmovision. In W. Jenkins, M. E. Tucker, & J. Grim (Eds), *Routledge handbook of religion and ecology* (pp. 158–168). Routledge.

Caruana, L. (2020). Different religions, different animal ethics? *Animal Frontiers*, 10 (1), 9–14. doi:10.1093/af/vfz047.

Dash, B., Narasimham, G. L., & Jinde, P. (2024). Animals in Hinduism: Exploring communication beyond the human realm in sacred texts and practices. *Journal of Dharma Studies*. doi:10.1007/s42240-024-00181-0.

DeMello, M. (2012). *Animals and society: An introduction to human–animal studies*. Columbia University Press.

Douglas, M. (1962). *Purity and danger: An analysis of concepts of pollution and taboo*. Routledge.

Dunn, P. M. (2006). Aristotle (384–322 BC): philosopher and scientist of ancient Greece. *Archives of Disease in Childhood (Fetal and Neonatal Edition)*, 91, F75–F77. doi:10.1136/adc.2005.074534.

Durkheim, E. (1912/2001). *The elementary forms of religious life*. Oxford University Press.

Fischer, J. (2023). *Vegetarianism, Meat and modernity in India*. Routledge. doi:10.4324/9781003319825.

Freidenreich, D. (2011). *Foreigners and their food: Constructing otherness in Jewish, Christian and Islamic Law*. University of California Press. doi:10.1525/9780520950276.

Govindrajan, R. (2018). *Animal intimacies: Interspecies relatedness in India's Central Himalayas*. University of Chicago Press. doi:10.7208/chicago/9780226560045.001.0001.

Halafoff, A., Harris, R., & Beaman, L. (2024a). (Non) religion, spirituality and the multispecies turn. *Social Compass*, 71(3), 389–405. doi:10.1177/00377686241276380.

Halafoff, A., Yunkaporta, T., Hancock, R., & Harris, R. (2024b). Riotous methodological and methodointuitive reflections on (non) religion, spirituality and the multispecies turn. *Social Compass*, 71(3), 482–501. doi:10.1177/00377686241276379.

Haraway, D. (2003). *The companion species manifesto: Dogs, people and significant otherness*. Prickly Paradigm Press.

Haraway, D. (2007). *When species meet*. University of Minnesota Press.

Haraway, D. (2016). *Staying with the trouble: Making kin in the Chthulucene*. Duke University Press. doi:10.2307/j.ctv11cw25q.

Harris, M. (1966). The cultural ecology of India's sacred cattle. *Current Anthropology*, 7(1), 51–59. doi:10.1086/200662.

Ingold, T. (2000). *The perception of the environment: Essays on livelihood, dwelling and skill*. Routledge.

Ip, E. C. (2025). Planetary health ethics: A Confucian alternative. *The Journal of Climate Change and Health*, 22, 100404. doi:10.1016/j.joclim.2024.100404.

Jalais, A. (2010). *Forest of tigers: People, politics and environment in the sundarbans*. Routledge India. doi:10.4324/9780203085516.

Kapek-Goodridge, A. (2024). Something fishy in the civilising process. In J. Lever, J. Smith Maguire, A. Kapek-Goodridge (Eds), *Towards an Eliasian understanding of food in the 21st century: Established foundations and new directions* (pp. 223–239). Springer. doi:10.1007/978-3-031-65774-0_12.

Kimmerer, R. W. (2013). *Braiding sweetgrass: Indigenous wisdom, scientific knowledge and the teachings of plants*. Penguin Books.
Kopnina, H., Roca, T., & Sorour, M. K. (2025). Integrating critical thinking about Sustainable Development Goals: Making biodiversity matter for business students. *Journal of Education for Sustainable Development*. doi:10.1177/09734082251336403.
Lever, J., & Fischer, J. (2018). *Religion, regulation, consumption: Globalising kosher and halal markets*. Manchester University Press. doi:10.7228/manchester/9781526103642.001.0001.
Levi Strauss, C. (1962). *The savage mind*. University of Chicago.
Linzey, A. (1987). *Christianity and the rights of animals*. Wipf and Stock Publishers.
Markowitz, F., & Avieli, N. (2022). Eating religiously: Food and faith in the 21st century. *Food, Culture & Society*, 25(4), 640–646. doi:10.1080/15528014.2022.2116195.
Masri, B. A. (2009). *Animals in Islam*. The Athene Trust.
McCance, D. (2013). *Critical animal studies: An introduction*. SUNY Press. doi:10.1353/book28954.
McLaughlin, R. P. (2012). Thomas Aquinas' eco-theological ethics of anthropocentric conservation. *Horizons*, 39(1), 69–97. doi:10.1017/S0360966900008549.
Midgley, M. (1984). *Animals and why they matter*. University of Georgia Press.
People for Animals v. State of Uttarakhand, Writ Petition No. 136 of 2010 (Uttarakhand High Court 2011).
Pope Francis. (2015). *Laudato Si': On care for our common home*. Vatican Press.
Robertson Smith, W. (1894). *Lectures on the religion of the Semites*. Adam and Charles Black.
Saha, S. (2024). Sacred serpents and the discourse on conservation: Interrogating interspecies dynamics in rural Bardhaman. *Sociologia Ruralis*, 64(2), 237–253. doi:10.1111/soru.12466.
Saha, S. (2025). Between the cultural and the ecological: Organic syncretism in Bardhaman's agrarian landscape. *South Asia: Journal of South Asian Studies*, 48(4), 782–803. doi:10.1080/00856401.2025.2526933.
Saha, S. (2026). Endangered cobras and conservation politics: Exploring multispecies encounters in agrarian landscapes of West Bengal. *Journal of Political Ecology* 33 (1), 6439. doi:10.2458/jpe.6439.
Scanes, C. G., & Chengzhong, P. (2018). Animals and religion, belief systems, symbolism and myth. In C. G. Scanes & P. Chengzhong (Eds) *Animals and human society* (pp. 257–280). Academic Press.
Shavarani, M. (2024). The relationship between two specific views in Islamic theology and environmental destruction. *International Journal of Social Science and Humanity*, 14(4), 182–185. doi:10.18178/ijssh.2024.14.4.1212.
Singh, B., & Dave, N. (2015). On the killing and killability of animals: Nonmoral thoughts for the anthropology of ethics. *Comparative Studies of South Asia, Africa and the Middle East*, 35(2), 232–245. doi:10.1215/1089201x-3139012.
Swanson, H. A., Lien, M. E., & Ween, G. B. (Eds). (2018). *Domestication gone wild: Politics and practices of multispecies relations*. Duke University Press. doi:10.1515/9780822371649.
Szűcs, E., Geers, R., Jezierski, T., Sossidou, E. N., & Broom, D. M. (2012). Animal welfare in different human cultures, traditions and religious faiths. *Asian-Australasian Journal of Animal Sciences*, 25(11), 1499–1506. doi:10.5713/ajas.2012.r.02.
Tallberg, L., García-Rosell, J. C., & Haanpää, M. (2022). Human–animal relations in business and society: Advancing the feminist interpretation of stakeholder theory. *Journal of Business Ethics*, 180(1), 1–16. doi:10.1007/s10551-021-04840-1.

Tsing, A. L. (2015). *The mushroom at the end of the world: On the possibility of life in capitalist ruins*. Princeton University Press. doi:10.1515/9781400873548.

Wade, R. (2004). Animal theology and ethical concerns. *Australian Ejournal of Theology*, 2. https://hdl.handle.net/20.500.14802/3832.

Weatherdon, M. S. (2022). Religion, animals, and indigenous traditions. *Religions*, 13 (7), 654, doi:10.3390/rel13070654.

Wilkie, R. (2005). Sentient commodities and productive paradoxes: The ambiguous nature of human–livestock relations in Northeast Scotland. *Journal of Rural Studies*, 21(2), 213–230. doi:10.1016/j.jrurstud.2004.10.002.

Wyller, T., Lundberg, J. G., & Gregersen, N. H. (2024). *Redeeming the sense of the universal*. Vandenhoeck & Ruprecht. doi:10.13109/9783666568749.

PART II

Animals as social and economic actors

4

ANIMALS AND CONSUMPTION

Introduction

Although the precise origins of the modern chicken are still debated, researchers agree that its primary ancestor was the Asian Red Junglefowl (*Gallus gallus*), first domesticated in Southeast Asia and Southwest China (Hata et al., 2021). Over millennia, chickens crossed continents alongside humans via trade routes, gradually becoming embedded in industrial food systems. The result is an animal that is today central to global biotechnological food production, its life reshaped by intensive genetic selection to maximise efficient muscle growth, feed optimisation, biosecurity regimes and planetary-scale metabolic flows (Barua, 2024). Drawing on Emma Roe's more-than-human geography of food systems, this chapter explores how industrial chicken consumption obscures the lives, labour and vulnerabilities of chicken-birds.

The rapid expansion of the chicken industry

In the last half century, global chicken production has expanded rapidly and by 2050 estimates suggest that it will have increased by more than 1,200% since 1960 (Barua, 2024; Fuseini et al., 2023). In the United Kingdom (UK) alone, over one billion broiler chickens (birds raised for meat rather than eggs) are consumed annually (Roe et al., 2024), making these birds the most consumed land animal in human history. Potts and Armstrong (2013) argue that these modern birds have been transformed from an animal valued for traits such as courage, broodiness and protective parenting into a figure of industrial meat production exploitation.

Examining industrial chicken production within a book on responsible management may seem contradictory at first glance. For some, the very act of managing animal consumption risks normalising systems of organised killing. This chapter

DOI: 10.4324/9781003585541-7

does not begin from the assumption that such systems are ethically settled. Rather, it argues that precisely because industrial food systems are structured through forms of multispecies vulnerability and violence, they demand critical engagement. Ignoring these systems does not reduce harm: It merely leaves their organisation, governance and ethical justification unexamined. An interspecies approach therefore requires confronting the contested realities of animal consumption.

Viewed as a "sentient commodity" (Wilkie, 2005), the "chicken-bird" (a term we use because "chicken," unlike other species, is also the name given to the meat harvested from their bodies) exemplifies how industrial food systems abstract and redesign animal life for economic value. The chicken-bird's body has been remade into one that grows rapidly, converting feed efficiently and meeting the growing global demand for cheap food and standardised protein (Miele, 2011; Striffler, 2007; Twine, 2024). Due to intensive selective breeding, by 2014 the bodyweight of broilers had more than quadrupled since the late 1950s, with 42-day birds weighing over 2.5 kg compared to around 0.6 kg in 1957. At the same time, the production cycle shortened dramatically, with broilers reaching market weight in half as much time as they once took to reach much lower weights (Zuidhof et al., 2014; see also Beldo, 2017).

This situation exemplifies the profound tensions between visibility and invisibility that shape contemporary consumption practices. Images of cartoon chickens through to breaded poultry products are everywhere, yet images that detail the broiler chicken's life and support its industrial life are nowhere. Visible on supermarket packaging, dietary guidelines, marketing campaigns and in cooking shows, the crowded sheds, imported soy-based feeds, antibiotic regimes and genetic vulnerabilities remain systematically concealed. This concealment makes chicken-birds a critical case for exploring the potential for a more visible, animal-centred approach to consumption grounded in vulnerability, relationality and ecological crisis. For managers operating within global food systems, these invisibilities are not peripheral but central to how responsibility, risk and sustainability are defined and governed. A *more-than-human* geographical perspective foregrounds the webs of metabolic, material and species interdependence that shape chicken-bird production and consumption of its meat (Barua, 2024), offering a route towards repairing fractured relations between humans and nonhumans in the food industry and beyond (Colombino & Bruckner, 2023; Gladkova & Matsuda, 2023). This chapter thus contributes to the book's broader project of developing an interspecies approach to responsible management by examining how industrial consumption reorganises multispecies life across global supply chains.

Contemporary chicken production and consumption

Contemporary sustainability discourses routinely position chicken as a climate-friendly success story capable of providing healthy, sustainable protein to feed a growing population (British Poultry Council, 2021; Tallberg et al.,

2024; Wang et al., 2010). Yet this narrative is profoundly anthropocentric. It presents chickens primarily as functional units of production whose value lies in their capacity to fulfil human economic and nutritional needs. By privileging human benefit, mainstream sustainability discourses obscure the harms experienced by chicken-birds and humans living and working in high-density production environments characterised by high security risks and zoonotic transmission (Tallberg et al., 2024). Despite the enormous scale of bird production, chickens' lives, deaths and distribution of parts of the chicken meat carcass remain largely absent from public discourse, policy and consumer consciousness. This invisibility is not incidental but actively produced through cultural norms, technological systems and economic structures that obscure interspecies entanglements and normalise systemic harm.

Norbert Elias's theoretical insights on the "process of civilisation" are useful for understanding this invisibility. Elias (1939/2012) documents how public acts of slaughter and butchery, which were once commonplace and witnessed openly in city streets, were gradually pushed out of everyday life. Historically, slaughter migrated from homes to specialised, hidden spaces on the urban periphery. This spatial reorganisation reshaped emotional dispositions, cultivating new standards of disgust, restraint and moral distancing. Industrialisation accelerated this cultural shift. Once part of everyday domestic life, the industrial broiler chicken has become a largely unknown and invisible component of industrial food systems (Striffler, 2007; Van Gelder & Dros, 2006). Such concealment does not merely hide production processes, it narrows the horizon of what can be perceived, imagined and politically addressed, ensuring that individualised consumption remains the dominant, unquestioned organising principle of the global food system.

Why a more-than-human perspective is useful

Addressing the consequences of concealment requires a *more-than-human* perspective that expands ethical and analytical attention beyond human needs and priorities. An animal-centred approach recognises chicken-birds as sentient beings with affective experiences, biological needs, social lives and behavioural repertoires (Marino, 2017). It also foregrounds their vulnerability within industrial systems that are designed not to enhance the wellbeing of animals, but to facilitate market optimisation, supply chain predictability and cost minimisation.

Tellingly, foregrounding chicken lives illuminates the hidden relationalities of production and consumption, relations that connect chicken meat to global feed markets, antibiotic manufacturers, river ecologies and distant forests cleared for soy production. From this perspective, chicken-birds are not simply victims of industrial intensification, they are central participants in multispecies metabolic systems. They consume vast quantities of soymeal, much of which is grown on deforested land in South America, particularly around the Amazon, as well as large quantities of antimicrobial drugs, vitamins, energy and water (Striffler, 2007; Twine, 2024; Van Gelder & Dros, 2006). Growing chicks metabolise these inputs, building up muscle, feathers, waste and emissions as they approach their slaughter-weight in a matter of weeks. Humans, in turn, consume the harvests (meat, feathers, bones) from their bodies along different product-chains, while the planet consumes their waste streams and the greenhouse gas emissions generated. This threefold consumption of chicken, human and planet reveals an unsustainable industrial system sustained by "metabolic labour" (Beldo, 2017).

Re-centring chicken-birds also widens the horizon of possible futures. Food system debates often focus narrowly on human security, hunger and agricultural efficiency. An animal-centred perspective invites alternative visions, lives worth living for chicken-birds, ecologies not forced to absorb harmful waste loads and multispecies vulnerabilities as shared conditions deserving of care. Making these relations visible expands the political imagination and opens up space for transformative change (Colombino & Bruckner, 2023; Gladkova & Matsuda, 2023).

Why broiler chickens are a useful case

Broiler chickens illuminate the deep contradictions of production and consumption in the ecological crisis. The vast scale of production positions chicken at the centre. Between 2010 and 2020, South America lost around 26,000 km^2 of forest annually, with global deforestation reaching 100,000 km^2 per year (Twine, 2024). Over the same period, global chicken populations increased by more than 30% to further drive land clearance and demand for soy-based feed (Van Gelder & Dros, 2006; Wangui et al., 2025).

These contradictions are also evident in local environmental crises in the UK. Alongside East Anglia, Lincolnshire and Norfolk, the Wye catchment on the England-Wales border has become a centre of industrial broiler production. The use of chicken waste as manure by local arable farmers has led to severe phosphorus pollution as the high-nutrient load has exceeded ecological limits (Barua, 2024). The associated river degradation and algal blooms reveal planetary systems working to metabolise agricultural waste. While chicken meat is framed as a sustainable healthy protein fit for human consumption (British Poultry Council, 2021), soy footprints, genetic fragility, antibiotic residues and nutrient runoff tell a different story. Rather than finding more sustainable solutions to harmful systems of chicken production and consumption, states and corporations increasingly govern such crises by monetising metabolic turbulence through feed optimisation, waste valorisation and other market based solutions (Barua, 2024). These dynamics expose not only ecological contradiction but governance failure, raising pressing questions for food retailers, supply-chain managers and regulators tasked with defining what counts as "sustainable" protein.

Our expert collaborator

Emma Roe, a leading more-than-human geographer at the University of Southampton, provides an invaluable empirical and theoretical basis for analysing these dynamics. Emma's research consistently foregrounds hidden *more-than-human* relations, examining how an ethics of care is positioned and often displaced within food production and consumption.

During her time working on the Welfare Quality® project at Cardiff University, Emma and her colleagues investigated how welfare certifications such as Red Tractor and RSPCA Assured cultivate the figure of the ethical consumer. As one of her colleagues noted, marketed as "happy" or "free-range" while their lives are reduced to technical metrics and audit scores (Miele, 2011), chicken-birds exemplify the contradictions and complexities of welfare politics and ethical consumption in markets (Miele & Lever, 2013). In *Food and Animal Welfare* (Buller & Roe, 2018), co-authored with Henry Buller, Emma developed the concept of "de-animalisation," exploring how animals become food through processes that strip away their sentience, materiality and sociality. Chicken meat may be metabolic matter, they argue, but it is matter that has mattered and perhaps should continue to matter.

As we discussed in our interview, Emma's subsequent chicken research stretches across global contexts. In Kenya and Malawi, she examined how WHO and Codex standards are negotiated locally, illuminating how global systems intersect with cultural practices, decolonial critiques and lived realities (Hughes et al., 2023). In the UK, she has worked with retailers and policy-makers on antimicrobial resistance, tracing how antibiotic governance shapes animal welfare, microbial ecologies and public health. Her current work conceptualises chickens as central nodes in an industrial socio-metabolic system, a system structured by feed imports, genetic selection, environmental degradation and multispecies vulnerabilities, which her work makes visible. By illuminating how chickens absorb, metabolise and embody global consumption patterns, this work contributes to an interspecies ethic grounded in shared ecological vulnerabilities and more-than-human interdependence.

What is consumption?

Scholarship on consumption has long been anchored in assumptions about individual choice and agency. Traditional sociological, anthropological and economic approaches frame consumption as a uniquely human domain in which identities are expressed, lifestyles created and preferences enacted (see Evans, 2019; Graeber, 2011). The individual consumer is positioned as a reflexive agent whose purchases signal ethical or political commitments, reinforcing a narrow understanding of consumption as something that happens at the point of purchase. Graeber (2011) suggests consumption operates ideologically, collapsing complex histories into a single, universalising moment.

Sustainable consumption scholarship is shaped similarly, with consumption often being imagined as a sphere of individual responsibility focused on selecting the "right" products rather than confronting the structural conditions of production. Scholars such as Evans et al. (2017) note how this "responsibilisation" obscures political–economic constraints that limit meaningful change. When it comes to engaging in ethical consumption (Crang & Hughes, 2015), consumers frequently evade direct responsibility for animal welfare, engaging in what has been termed "voluntary ignorance" (Harper & Henson, 2001, as cited in Van Beirendonck et al., 2013) or by delegating responsibility to farmers, retailers and regulatory authorities. The expansion of market-oriented welfare initiatives through labels and brand-based assurance schemes conceals as much as it discloses, protecting retailers and food companies from critical public scrutiny (Buller & Roe, 2018; Freidberg, 2003; Miele, 2011).

Elias's (1939/2012) analysis of the "process of civilisation" illuminates how violence, slaughter and other troubling aspects of food production have gradually been hidden from public life. This spatial and emotional distancing came to underpin contemporary consumption, with animals appearing as products rather than beings and consumers shielded from the visceral realities of production – a dynamic that Lever (2026) argues is central to modern food systems. Building on this insight, Twine (2024) suggests that the slaughterhouse now functions as a cultural secret whose invisibility protects anthropocentric norms from critical scrutiny, shielding children from the realities of industrial production and potentially shaming parents. Such concealment narrows what can be perceived, imagined, or ethically contested, ensuring that the food system remains organised around individual consumption rather than collective responsibility.

Other ecofeminist and critical animal studies scholars have interrogated the anthropocentric nature of consumption. Adams' (1990) concept of the "absent referent" describes how linguistic, visual and material forms strip animals of agency and identity, enabling their transformation into consumable objects. This absence is not merely symbolic, it structures how consumers understand food, effacing the lives, deaths and multispecies relations involved. Colombino and Bruckner's (2023) notion of "anthroparchy" (explained below) captures this broader organisation of power that systematically marginalises animal lives while centring human desires, comforts and narratives.

Emerging multispecies and more-than-human approaches seek to challenge this paradigm by repositioning consumption as a relational, ecological and metabolic process involving humans, animals, technologies and planetary systems. Rather than treating consumption as an isolated human act, these frameworks foreground the interdependencies and industrial exploitation of vulnerabilities that underpin production and consumption across multispecies communities. Indeed, vulnerable human consumers are just as susceptible to exploitation and its consequences as the chickens whose lives are produced and consumed.

While Elias' (1939/2012) reflections on civilisation show how individualisation weakened collective "we-identities," enabling private consumption to dominate, Lever (2026) argues that today's socio-ecological crises unsettle this model, pushing more relational understandings. Haraway's notion of "sympoiesis" – making with – reinforces this shift by showing that all "we-images" are multispecies, with humans and chickens co-creating ecological worlds through shared metabolisms (Haraway, 2008, 2016). Bringing Elias and Haraway together reframes consumption, demonstrating that humans, chickens and the planet share metabolic processes, vulnerabilities and futures. As Roe's work demonstrates, what we call "consumption" is always multispecies. This reframing destabilises anthropocentric assumptions and opens pathways for rethinking responsibility, ethics and governance beyond the consumer-centric model.

Towards a visible more-than-human politics of consumption

Chicken-birds are the most consumed land animal in human history, yet today they are both hyper-visible and profoundly concealed. Chicken meat populates supermarket shelves, advertisements and dietary guidelines as symbols of lean, sustainable protein fit for human consumption (British Poultry Council, 2021; Tallberg et al., 2024; Wang et al., 2010). Yet the actual lives of chicken-birds – the crowded sheds, rapid-growth genetics, antibiotic dependence and mass slaughter – remain largely invisible, hidden behind the everyday familiarity of everyday food products.

This concealment underpins a historical "process of civilisation" (Elias, 1939/2012) through which modern societies sanitise or hide practices that expose their violent foundations. Just as slaughterhouses were moved from city centres to urban peripheries in the 19th century, so the infrastructures and ecologies of contemporary industrial broiler production have been pushed out of sight and mind (Lever, 2026; Striffler, 2007). For chicken products to be comfortably consumed, they must be made emotionally unobjectionable (Adams, 1990; Grauerholz, 2007) – hence the huge popularity of the chicken nugget (Patel & Moore, 2018). Managers often deploy moral insulation strategies to distance consumers (and themselves) from ethical implications, normalising problematic practices through moral dissonance (Lowell, 2012). In patriarchal cultures, Adams (1990) argues that animals, like women, become objectified, symbolically erased and visually transformed in ways that legitimise their exploitation. This erasure is achieved through the removal of the living animal, the linguistic renaming of bodies as food and their fragmentation into cuts or processed forms that no longer resemble a creature.

Emma Roe's latest research on chicken, metabolism and the socio-material flows of industrial food systems invites a sharper focus on consumption, not only the consumption of different parts of the chicken by different communities of humans, but also chicken-birds' consumption of soy, dietary supplements, antibiotics and vaccinations and the planet's consumption of the waste, emissions and pollution. In Roe's work, visibility becomes a method for revealing the metabolic entanglements that shape contemporary food systems, connecting chicken-birds, humans, microbes, farmers and rivers.

Making consumption visible

To "include" animals analytically in this way is to confront how their lives and deaths underpin human consumption practices. Yet as Colombino and Bruckner (2023) show, animals remain materially central and symbolically marginal in the objects that we eat, wear and use and are thus largely absent from thought (Adams, 1990). The organising principle *"anthroparchy"* refers to the structural dominance of the human that relegates other beings to the periphery of moral concern and political visibility. However, invisibility is not simply about a lack of attention, it is also about governance. What is not seen cannot be cared for, regulated or considered ethically.

During our interview, Emma noted that chicken consumption is not a simple act of ingestion, but a key point in planetary metabolic flows. The broiler chicken cycle of production is itself a voracious consumer of resources, dependent on the complex global supply chains that sustain the birds' rapid muscle growth. Among other things, chicken-birds consume huge amounts of imported soymeal, antimicrobial drugs and industrial-scale quantities of water and energy (Barua, 2024; Van Gelder & Dros, 2006). Their bodies metabolise these inputs into muscle, bones and waste, which are then harvested to supply various consumers – from low-income shoppers wanting cheap, pre-cooked chicken wings through to the arable farmer seeking an inexpensive source of manure for his crops – while the planet absorbs the waste streams produced. This threefold consumption of chicken, human and planet underpins industrial chicken production.

Through the metabolic mapping carried out by Emma's team, this hidden consumption at the heart of this industrial production system becomes visible:

- Soymeal from South America grown on deforested land feeds British chickens.
- Antibiotics control the spread of disease pathogens between chickens living in confined sheds but accelerate antimicrobial resistance across species and environments.
- Carcass parts travel globally, with desirable cuts consumed domestically and unwanted cuts shipped to Asia and Africa, or turned into a processed animal product.
- Phosphorus and nitrogen runs off land into waterways, contributing to river degradation and ecological collapse.

Although they remain invisible to most consumers, these flows constitute the infrastructure of industrial chicken production. As Barua (2024) argues, metabolisms, both human and nonhuman, are intertwined through global circulations of matter, making consumption not just a personal preference, but a socio-ecological event with far-reaching consequences.

Consumption and concealment

A profound form of concealment lies in chicken feed, especially soy. Modern poultry systems rely on vast quantities of soymeal imported from the Americas, much of it cultivated on land cleared of forests, wetlands and indigenous territories. Soy is the primary protein source in broiler diets that underpins the rapid growth of the sector, yet soy expansion is also a key driver of deforestation, biodiversity loss and greenhouse gas emissions (Twine, 2024; Van Gelder & Dros, 2006). The industry's over-reliance on soy is unsustainable, but while its ecological footprint is rendered invisible, soy is often marketed as a clean, plant-based ingredient (Wangui et al., 2025). Its incorporation into broiler feed quietly binds consumers to distant landscapes of extraction, even as chicken meat is marketed as "sustainable," "high welfare" or from "happy chickens" (Baumann et al., 2025; Miele, 2011; Miele & Lever, 2013). In this sense, consumers not only ingest chicken flesh, they also consume the metabolic history of the soy, energy, water and antibiotics that have sustained the chicken's life.

To treat and prevent disease in ever larger populations of industrially farmed broilers, feed includes antibiotics (Beldọ, 2017). In some global regions this prophylactic (preventative) treatment of broilers is banned (Ryan, 2018), while in other places such use of antibiotics is still common (Hughes et al., 2023). Chickens consume these drugs and if there is insufficient withdrawal time before slaughter and human consumption, humans consume their residues and the planet consumes the human sewage. In this sense, mis-handled antibiotics are potentially consumed three times by chickens, by humans and the planet. Emma's mapping positions antibiotic circulation as a critical dimension of global chicken meat consumption, one that makes visible the multispecies consequences of production and its contributions to antimicrobial resistance.

Chickens as consumers and consumed

Unlike some other farmed species, we noted above that we use the same word to describe the meat ("chicken") that is harvested for consumption and the living being ("chicken") that is itself a consumer. Seeing chickens in this way, as consumed and consumer (Hurn, 2016), highlights how their feeding practices actively shape global ecologies by metabolising vast flows of soymeal, antibiotics, minerals, water and energy. The human management of their appetites drives deforestation, soil depletion, pharmaceutical circulation and nutrient pollution. Significantly, this positions chicken-birds as metabolic agents whose needs restructure landscapes, supply chains and multispecies relations. Understanding them in this way, as consumers, is important, as it exposes consumption as a multispecies world-making process extending beyond the human world and its anthropocentric assumptions.

This dual role also explains the "double invisibility" of industrial chicken. While the inputs chicken-birds consume are hidden from view, their lives and suffering are simultaneously concealed behind cheap, abundant chicken meat products. The illusion of sustainability depends on ignoring both sides of this metabolic relation. Acknowledging that chickens eat, metabolise and suffer dissolves the human–animal boundary protecting consumer innocence. If chicken-birds are consumers like us, eating them becomes a relational entanglement. Corporations work to manage this discomfort through narratives of welfare, sustainability and choice (Baumann et al., 2025), yet visibility disrupts these narratives by refusing the sanitised simplicity of the supermarket label underpinning individualised consumption patterns.

Carcass balance and the uneven geography of consumption

Roe et al.'s (2024) insights into carcass balance shine a light on the "civilised" realities of ethical and sustainable consumption practices. Consumers in the UK prefer breast meat, treating it as the most desirable and healthy part of the bird's body. While chicken meat may be labelled "high welfare," "healthy" and "sustainable," these preferences drive breeding practices that distort the bodies of chicken-birds, resulting in health issues such as lameness and cardiovascular strain.

Meanwhile, the less desired cuts such as legs, wings and feet are exported to other markets, particularly in Asia and Africa. As Emma notes during our interview:

> The ethical consumer buying an organic breast believes they're making the best choice. But they're supporting the bringing into life of an entire animal. Others then have to deal with the rest of that carcass.

Trade patterns confirm this asymmetry. In 2022, the UK imported over 500,000 tonnes of white poultry meat while exporting 254,000 tonnes of darker cuts (Roe et al., 2024). Consumption thus redistributes not only animal parts but ethical responsibilities and ecological consequences across global markets. The ethical consumer's choice remains parasitic on the metabolic labour of others – human and nonhuman.

Environmental consequences of the planet as consumer

Chicken production generates waste that the planet is forced to consume. In the Wye Valley, as we have seen, phosphorus runoff from arable farms that use poultry manure has contributed to severe eutrophication, algal blooms and ecological collapse (Barua, 2024). Nutrient cycles once mediated by diverse ecosystems are now overloaded by the metabolic outputs of concentrated animal feeding operations. Antibiotic residues seep into soils and waterways, while ammonia emissions contribute to air pollution and climate change.

Emma insists that these environmental impacts cannot be separated from welfare or public health. Broilers, rivers, microbes and humans are metabolically bound; the planet absorbs the waste generated by industrial agriculture, even as marketing frames chicken meat as a sustainable choice. Seeing the planet as a consuming entity highlights asymmetries of power and the ways in which ecosystems ingest the byproducts of human consumption without consent. In this way, visibility becomes a tool for reframing responsibility. It shifts the burden from individual consumers, whose choices are shaped by limited availability of information on the infrastructures, corporations and regulatory regimes that organise and conceal these flows.

Rethinking consumption through a more-than-human politics of visibility

Emma's concept of the "industrial socio-metabolics of chicken" reframes consumption as a multispecies process shaped by global supply chains and ecological limits. Industrial chicken production becomes a vast metabolic system that consumes chickens' bodies, soy, antibiotics and energy. This holistic framing reveals consumption as planetary and relational rather than individual or discrete. Read alongside Gladkova and Matsuda's (2023) speculative work reimagining pigs as collaborators rather than commodities, the tracing of material flows in Emma's research offers a grounded counterpoint. Both approaches challenge consumer-centric ethics and reposition animals as central to food futures.

Barua (2024) argues similarly that metabolism is a distinctive way of diagnosing how life is managed under capitalism, foregrounding movement, transformation and disruption across heterogeneous assemblages. Through this lens, chicken-birds are not simply consumed; they are metabolically entangled with microbes, farmers, rivers and consumers. Making these entanglements visible shifts responsibility from individual shoppers to infrastructures where feed, welfare, pollution and antibiotic governance intersect.

Visibility also reconfigures governance by demanding multispecies accountability across feed, welfare, pollution and trade infrastructures. It transforms care by extending ethical concern beyond labels to soil, water, microbes and distant communities. Finally, as Tsing (2015) notes, viewing multispecies relations is a precondition for imagining alternative futures. While individualised consumption patterns are premised on an overarching sense of "*we-lessness*" (Elias, 2010), today's socio-ecological crises are intensifying calls for more relational forms of "*we-ness*" (Haraway, 2008, 2016; Lever, 2026) shaped by recognition of shared multispecies metabolisms.

Directions for future research

Future research on "chicken," "consumption" and "food systems" must extend the multispecies, socio-metabolic and visibility-focused agenda outlined in this chapter. The industrial chicken offers a uniquely generative site for deepening debates on sustainability, ethics, governance and relationality and taking this work forward requires innovation across several fronts.

A more-than-human politics of visibility depends on methodological tools capable of following flows, relationships and concealments across multiple scales. Emma's socio-metabolic mapping shows how soymeal, antibiotics, water, energy, carcass parts and waste circulate through industrial systems in ways that are constitutive of animal, human and ecological vulnerability (Barua, 2024; Roe et al., 2024). Further methodological innovation is needed to make industrial practices more visible, accountable and open to political challenge. Ethnographic work could be combined with visualisation technologies, sensor data and participatory mapping to reveal hidden processes that escape public and policy attention. Speculative and arts-based methods, such as those used by Gladkova and Matsuda (2023) to reimagine pig–human relationships, also offer routes for unsettling anthropocentric assumptions and opening imaginative space for thinking with chickens as co-participants rather than commodities.

Because many studies remain context-specific, future work should also take a comparative perspective. The redistribution of carcass parts, that is, domestic demand for breast meat coupled with exports of darker cuts to markets in Africa and Asia (Roe et al., 2024; Striffler, 2007), shows how cultural preferences, trade relations and ethical responsibilities intersect. Emma's research in Kenya and Malawi similarly demonstrates how WHO and Codex standards are translated differently through local histories, infrastructures and cultural values, highlighting the importance of decolonial perspectives that challenge universalist framings of "sentience," "welfare" and "expertise." Comparative studies could therefore illuminate how visibility and concealment are produced differently across political, economic and cultural settings.

A further area for development lies in interrogating existing governance models. Although frameworks such as One Health emphasise connections between human, animal and ecological wellbeing (Deem et al., 2019), they remain fundamentally anthropocentric, rarely treating nonhuman animals as stakeholders in their own right (Tallberg et al., 2024). Future research should explore governance models that centre chicken-birds as vulnerable actors embedded in socio-ecological systems. This includes critical engagement with certification schemes, which often function as pacifiers that obscure systemic harms behind welfare labels and metrics (Buller & Roe, 2018; Miele, 2011). It also involves examining how states and corporations govern metabolic turbulence through feed optimisation, waste valorisation and pollution-credit systems, mechanisms that, as Barua (2024) argues, regulate crises in ways that open new avenues for profit rather than addressing underlying harms.

Finally, there is a need for research that examines how invisibility is reproduced pedagogically and epistemically. The absence of chicken-birds from public consciousness is not only a material condition, but an educational and cultural one. Future research might investigate how classroom teaching, curriculum design and public engagement can cultivate multispecies literacy and support the development of ethical imagination (Tsing, 2015). This includes attending to the affective dynamics of discomfort, avoidance and shame that shape public responses to animal suffering and environmental harm. Exploring how pedagogical interventions might transform these responses into ethical and political attentiveness would extend multispecies scholarship into everyday life and support a wider cultural shift towards recognising chicken-birds as sentient beings embedded in socio-metabolic systems.

Ethical implications

An interspecies approach does not presume the ethical legitimacy of animal killing, nor does it assume that reforming industrial systems is sufficient. Rather, it begins from the recognition that contemporary food systems are structured through forms of organised violence that demand ethical scrutiny. The aim is not to sanitise these realities, but to render them visible, contestable and accountable within governance and management practice.

Making the realities of chicken meat production visible has profound ethical implications. Industrial poultry systems rely on forms of concealment that obscure animal suffering, ecological degradation and public health risks, enabling these practices to continue without moral disruption. Emma's work on "de-animalisation" highlights how animals are transformed into units of "production" or "protein" that erases their sentience and materiality (Buller & Roe, 2018). When chicken-birds are treated as abstractions rather than beings with capacities, needs and vulnerabilities, the moral legitimacy of their systematic exploitation is rarely questioned. Visibility challenges this reduction by insisting on recognition of these birds as more-than-economic actors, prompting uncomfortable questions about the ethics of treating them as disposable life.

The ethical implications extend to consumer responsibility. Welfare labelling schemes and sustainability claims often function as pacifiers, offering moral reassurance (Buller & Roe, 2018; Miele, 2011) while obscuring stocking densities, genetic manipulation, toxic waste flows and antibiotic dependence (Barua, 2024). These schemes conceal systemic harms beneath familiar symbols of "care," enabling consumers to feel ethical without engaging with the structural conditions that shape chicken-birds' lives. Making these contradictions visible destabilises narratives of "ethical choice" and reveals the limits of consumer-driven moral action. If consumers are reassured that the market can "fix" welfare, they are less likely to question the deeper logics of intensification that generate harm in the first place.

Visibility also reveals uneven geographies of ethical responsibility. Carcass redistribution, where breast meat is retained in the Global North while darker cuts, feet and wings are exported South, creates ethical asymmetries that are rarely acknowledged (Roe et al., 2024; Striffler, 2007). Similarly, the global flows of soymeal from deforested regions of South America, which underpin the rapid growth of individual chickens and the poultry industry, tie everyday chicken meat consumption to distant ecologies and communities (Twine, 2024; Van Gelder & Dros, 2006). A decolonial lens further shows how the burdens of production, pollution and waste disproportionately fall on certain landscapes and populations, while economic benefits accrue elsewhere. Ethical debate cannot therefore remain confined to individual consumer actions or industrial food system debates about sustainability, it must encompass the globally unequal infrastructures that support chicken production.

Ultimately, a multispecies ethic of care (Haraway, 2016; Puig de la Bellacasa, 2017; Tsing, 2015) demands a shift beyond anthropocentric frameworks. If chicken-birds are recognised as sentient and vulnerable actors entangled in socio-metabolic systems, the question is no longer simply how to minimise harm, but how to reconceive food systems in ways that honour their lives and question unequal socio-ecological relations. This requires attending to the political, cultural and affective conditions that determine what is seen and what is rendered invisible. Ethical responsibility becomes collective rather than individual, grounded in an understanding of shared vulnerability across human, animal and ecological worlds. Visibility, in this sense, is not simply an act of exposure, but a precondition for imagining more just multispecies futures.

Practical implications for responsible management

A more-than-human politics of visibility has significant consequences for how responsibility is understood and enacted in food system management. Retailers, producers, certification bodies and policymakers have traditionally relied on consumer-facing indicators, especially welfare labels, to demonstrate accountability. Yet these schemes frequently pacify consumers by translating complex welfare and ecological concerns into logos, scores and simplified narratives (Buller & Roe, 2018; Miele, 2011). Responsible management requires moving beyond symbolic reassurance towards transparent disclosure of stocking densities, feed sourcing, antibiotic regimes, energy demands and ecological impacts. Such transparency is not merely ethical but strategic. Obscured harms, such as phosphorus pollution in the River Wye (Barua, 2024), will eventually manifest as reputational, regulatory and ecological crises that undermine long-term resilience.

As Emma's work recognises, resilience itself must be redefined. Typically conceptualised as supply-chain continuity or operational stability, resilience in the poultry sector cannot be disentangled from animal wellbeing and ecological limits. Roe's work suggests that industrial fragility brought about through overcrowding, rapid-growth genetics and dependence on antibiotics creates systemic vulnerabilities that no amount of managerial optimisation can resolve. Practically, this means embedding welfare and environmental thresholds into scenario planning and risk governance to prevent socio-ecological fallout. It also demands confronting the structural drivers of antimicrobial resistance rather than relying solely on downstream mitigation.

Global interdependencies further complicate responsible management. The industry's reliance on soy imported from deforested regions (Twine, 2024; Van Gelder & Dros, 2006) links corporate decision-making to ecological harm in distant landscapes. Carcass redistribution reinforces ethical inequalities, with surplus cuts exported to markets in the Global South (Roe et al., 2024; Striffler, 2007). Responsible management must therefore address these asymmetries by diversifying feed sources, collaborating on deforestation-free supply chains and supporting equitable trade relationships that do not externalise environmental and ethical burdens.

To be truly effective, responsible management requires recognising chickens as stakeholders in their own right in multispecies systems (Tallberg et al., 2024). This means designing governance frameworks that integrate ecological health, animal welfare and human wellbeing and developing assurance schemes that measure multispecies outcomes, including soil-nutrient balance, water quality, biodiversity indicators and welfare metrics, rather than relying on consumer-facing moral indicators. By outlining these issues, this chapter repositions chicken-birds not as units of production but as central actors in sustainable and responsible food-system governance.

Reflective activities inside/outside the classroom

Students can deepen their understanding of multispecies consumption in the classroom by engaging with reflective questions and activities that make invisibilities visible; outside the classroom, they can engage in sensory or place-based reflection. Both types of activity help them to link abstract knowledge to lived environments and bodies, cultivating attentiveness, provoking ethical questioning and supporting a shift from individualised moral narratives to systemic and multispecies understandings of responsibility.

A Revision questions that test knowledge ("what?")

- A1 What dominant economic and ethical framings does the concept of sentient commodities draw attention to?
- A2 What has the process of civilisation concept contributed to our understanding of the industrial chicken production and consumption?
- A3 What do welfare labels reassure and what do they conceal?
- A4 How does the chapter describe the "threefold consumption" of chickens, humans and the planet within industrial food systems?

B Reflective questions that build understanding ("so what?")

- B1 In what ways are chickens' lives hidden from the public?
- B2 How does the invisibility of soy reshape our understandings of consumption and sustainability?
- B3 How and in what ways do states and corporations govern environmental crises unsustainably?
- B4 What changes when consumption is understood as a multispecies metabolic process rather than an individual consumer choice?

C Take action ("do")

- C1 Map the supply-chain journey of a chicken product, identifying hidden flows of feed, labour, energy and waste.
- C2 Analyse chicken representations in supermarkets or takeaway outlets, noting what is shown and what is obscured.
- C3 Keep a personal diary of encounters with chicken over one week, documenting presence, absence and emotional responses.
- C4 Trace the origin of soy used in UK poultry feed and investigate its ecological impacts.

D Ideas to discuss with colleagues, friends and family ("share")

- D1 Debate whether ethical consumption can meaningfully address systemic harms or whether governance must shift beyond markets.
- D2 Reflect on how childhood exposure or shielding from the realities of meat production shapes adult attitudes towards the climate crisis.
- D3 Explore the River Wye as an example of how ecological damage is entangled with everyday meals.
- D4 Consider what "responsibility" might look like if chickens were treated as stakeholders in food systems.

Key ideas

Absent referent

A concept developed to explain how animals are linguistically and materially converted into products such as meat, thereby obscuring the reality of their lives and deaths. The term captures the process through which the original subject disappears from view in the finished commodity, rendering their existence invisible within systems of consumption (Adams, 1990).

Carcass asymmetries

Consumer demand for breast meat generates significant imbalances within the global poultry industry. As a result, surplus darker cuts are frequently exported to countries in the Global South, effectively shifting ecological burdens and ethical consequences elsewhere. This illustrates how routine consumption practices are embedded within global trade structures that contribute to broader forms of multispecies injustice (Striffler, 2007).

Multispecies resilience

Industrial poultry production erodes both ecological and animal resilience through the use of rapid-growth breeding regimes, routine antibiotic inputs and the accumulation of nutrient pollution. Resilience can instead be understood as a multispecies condition that depends not on market efficiency or stability, but on the wellbeing of animals, the integrity of microbial ecologies and the maintenance of ecological limits.

The politics of visibility

Civilised concealment shields anthropocentric norms from sustained critical examination by obscuring the violence and labour embedded in animal production systems. A politics of visibility seeks to unsettle this blind spot by reconnecting consumers with the material realities of hidden labour, animal suffering and multispecies injustice. In so doing, it reframes responsibility as systemic rather than individual, directing attention towards the institutions, regulatory frameworks and corporate actors that structure these conditions (Twine, 2024).

Sentient commodities

When chicken-birds are framed primarily as commodities, their sentience is strategically obscured within market logics that prioritise efficiency, yield and profit. Truly recognising these birds as sentient beings unsettles the assumption that market mechanisms alone can resolve our ethical obligations. Such recognition exposes the limits of commodity production and calls for moral and political frameworks that take animal subjectivity seriously (Marino, 2017; Wilkie, 2005).

References

Adams, C. J. (1990). *The sexual politics of meat: A feminist-vegetarian critical theory*. Continuum.

Barua, M. (2024). Metabolic politics: A comparative synthesis. *Transactions of the Institute of British Geographers*, 50(2), e12712. doi:10.1111/tran.12712.

Baumann, S., Kennedy, E. H., Johnston, J., & Oleschuk, M. (2025). *Happy meat: The sadness and joy of a paradoxical idea*. Stanford University Press.

Beldo, L. (2017). Metabolic labor: Broiler chickens and the exploitation of vitality. *Environmental Humanities*, 9(1), 108–128. doi:10.1215/22011919-3829154.

British Poultry Council. (2021). *Let's redefine sustainability*. https://britishpoultry.org.uk/lets-redefine-sustainability.

Buller, H., & Roe, E. (2018). *Food and animal welfare*. Bloomsbury. doi:10.5040/9781474204330.

Colombino, A., & Bruckner, H. K. (2023). Hidden in plain sight: How (and why) to attend to the animal in human–animal relations. In A. Colombino & H. K. Bruckner (Eds), *Methods in human–animal studies* (pp. 1–29). Routledge. doi:10.4324/9781351018623-1.

Crang, M., & Hughes, A. (2015). Globalizing ethical consumption. *Geoforum*, 67. doi:10.1016/j.geoforum.2015.10.005.

Deem, S., Lane-de Graaf, K., & Rayhel, E. (2019). *Introduction to One Health: An interdisciplinary approach to planetary health*. Wiley Blackwell.

Elias, N. (1939/2012). *On the process of civilisation*. UCD Press.

Elias, N. (2010). *The society of individuals (collected works)*. Bloomsbury.

Evans, D., Welch, D., & Swaffield, J. (2017). Constructing and mobilising "the consumer." *Environment and Planning A*, 49(6), 1396–1412. doi:10.1177/0308518X17694030.

Evans, D. M. (2019). What is consumption, where has it been going and does it still matter? *The Sociological Review*, 67(3), 499–517. doi:10.1177/0038026118764028.

Freidberg, S. (2003). Cleaning up down South: Supermarkets, ethical trade and African horticulture. *Social and Cultural Geography*, 4(1), 27–43. doi:10.1080/1464936032000049298.

Fuseini, A., Miele, M., & Lever, J. (2023). Poultry welfare at slaughter. *Poultry*, 2(1), 98–110. doi:10.3390/poultry2010010.

Gladkova, E., & Matsuda, N. (2023). Reimagining pigs – A multispecies, ecofeminist research method. *Australian Feminist Studies*, 38(115–116), 177–197. doi:10.1080/08164649.2024.2309619.

Graeber, D. (2011). Consumption. *Current Anthropology*, 52(4), 489–511. doi:10.1086/660166.

Grauerholz, L. (2007). Cute enough to eat: The transformation of animals into meat for human consumption in commercialized images. *Humanity and Society*, 31(4), 334–354. doi:10.1177/016059760703100404.

Haraway, D. J. (2008). *When species meet*. University of Minnesota Press.

Haraway, D. J. (2016). *Staying with the trouble*. Duke University Press. doi:10.2307/j.ctv11cw25q.
Hata, A., Nunome, M., Suwanasopee, T. (2021). Origin and evolutionary history of domestic chickens. *Scientific Reports*, 11(1), 2035. doi:10.1038/s41598-021-81589-7.
Hughes, A., Roe, E., Wambiya, E. (2023). The challenges of implementing antibiotic stewardship in diverse poultry value chains in Kenya. *Agriculture and Human Values*, 41, 749–767. doi:10.1007/s10460-023-10518-3.
Hurn, S. (2016). Animals as producers, consumers and consumed: The complexities of trans-species sustenance in a multi-faith community. *Ethnos*. doi:10.1080/00141844.2015.1107611.
Lever, J. (2026). Established foundations and new directions for an Eliasian sociology of food. In A. Murcott & M. Niva (Eds), *Routledge international handbook of the sociologies of food and drink* (pp. 83–97). Routledge.
Lowell, J. (2012). Managers and moral dissonance: Self-justification as a big threat to ethical management? *Journal of Business Ethics*, 105, 17–25. doi:10.1007/s10551-011-0931-9.
Marino, L. (2017). The inconvenient truth about thinking chickens. *Animal Sentience*, 2(17), 1. doi:10.51291/2377-7478.1211.
Miele, M. (2011). The taste of happiness: Free-range chicken. *Environment and Planning A*, 43(9), 2076–2090. doi:10.1068/a43257.
Miele, M., & Lever, J. (2013). Civilising the market for welfare friendly products in Europe? The techno-ethics of the Welfare Quality® assessment. *Geoforum*, 48, 63–72. doi:10.1016/j.geoforum.2013.04.003.
Patel, R., & Moore, J. (2018). How the chicken nugget became the true symbol of our era. *The Guardian*, 8 May.www.theguardian.com/news/2018/may/08/how-the-chicken-nugget-became-the-true-symbol-of-our-era.
Potts, A., & Armstrong, P. (2013). Picturing cruelty; chicken advocacy and visual culture. In J. Johnston & F. Probyn-Rapsey (Eds), *Animal death* (pp. 151–168). Sydney University Press.
Puig de la Bellacasa, M. (2017). *Matters of care: Speculative ethics in more-than-human worlds*. University of Minnesota Press.
Roe, E., Maye, D., Lambton, S. (2024). *Changing agri-chicken for net zero*. University of Southampton.
Ryan, C. (2018). EU bans prophylactic use of antibiotics in farming from 2022. *Poultry News*, October 26. www.poultrynews.co.uk/news/eu-bans-prophylactic-use-of-antibiotics-in-farming-from-2022.html.
Striffler, S. (2007). *Chicken: The dangerous transformation of America's favourite food*. Yale University Press.
Tallberg, L., Huopalainen, A., & García-Rosell, J. C. (2024). Beyond anthropocentrism: A call to action for multispecies inclusivity. *Business & Society*. doi:10.1177/00076503241271254.
Tsing, A. L. (2015). *The mushroom at the end of the world*. Princeton University Press. doi:10.1515/9781400873548.
Twine, R. (2024). *The climate crisis and other animals*. Sydney University Press. doi:10.30722/sup.9781743328996.
Van Beirendonck, S., Driessen, B., & Geers, R. (2013). Belgian consumers' opinion on pork consumption concerning alternatives for unanesthetized piglet castration. *Journal of Agricultural and Environmental Ethics*, 26(1), 259–272. doi:10.1007/s10806-012-9378-y.

Van Gelder, J. W., & Dros, J. M. (2006). *From rainforest to chicken breast: Effects of soybean cultivation for animal feed on people and nature in the amazon region*. Friends of the Earth Netherlands and Cordaid.

Wang, Y., Lehane, C., Ghebremeskel, K., & Crawford, M. A. (2010). Modern organic and broiler chickens sold for human consumption provide more energy from fat than protein. *Public Health Nutrition*, 13(3), 400–408. doi:10.1017/S1368980009991157.

Wangui, C., Taylor, J., Barraza, H., Hancock, P., & Kyriazakis, I. (2025). One size does not fit all: Are there any sustainable alternatives to soybean in chicken systems? *Poultry Science*, 105957.

Wilkie, R. (2005). Sentient commodities. *Journal of Rural Studies*, 21(2), 213–230. doi:10.1016/j.jrurstud.2004.10.002.

Zuidhof, M. J., Schneider, B. L., Carney, V. L., Korver, D. R., & Robinson, F. E. (2014). Growth, efficiency, and yield of commercial broilers from 1957, 1978, and 2005. *Poultry Science*, 93, 2970–2982.

5

ANIMALS AND WORK

Introduction

Animals perform all kinds of work in all kinds of ways: From the dogs sniffing out drugs at a British airport, to the donkeys carrying vegetables to an Ethiopian market and the birds spreading seeds across an Amazonian forest glade. Yet when we talk of workers, we don't usually imagine that they might have paws, hooves or beaks.

The concept of work has traditionally been tied to economic productivity and human activity. Critical perspectives (e.g. Glucksmann, 1995) have broadened the scope to include unpaid and informal labour. However, given that we share our communities and organisations with other nonhuman animals, these frameworks remain incomplete unless we account for the significant contribution of animals as well. Including animals in discussions about work therefore challenges the anthropocentric biases inherent in our thinking and behaviour, enabling a fuller understanding of work as a life-sustaining, relational activity shaped by diverse forms of collaboration and labour that cut across the boundaries between humans and other animals.

Drawing on Kate Dashper's work on human-animal relations in equestrian sport, leisure and tourism, this chapter explores what it means to understand work as an interspecies endeavour and to recognise animals as co-workers.

The value of an animal-centred approach to work

There are at least three interrelated justifications for focusing on animal work. From a practical point of view, whether we acknowledge it or not, humans and our organisations are effectively mixed up with animals of all kinds (Donaldson & Kymlicka, 2011; Tallberg & Hamilton, 2022). From a

DOI: 10.4324/9781003585541-8

moral perspective, then, we owe it to other species to take their work seriously, especially when the animals in question often have little choice as to when, where and how this work is undertaken. Recognising animals as workers requires a reconsideration of our ethical obligations towards them (Blattner et al., 2020). Likewise, from an academic perspective, animals are integral to many organisational processes, contributing physically, emotionally and cognitively to tasks and outcomes. Ignoring these contributions perpetuates a narrow and exclusionary view of work but also limits our understanding of the relational, ethical and ecological dimensions of organisational life (Kandel et al., 2023; Tallberg et al., 2022).

By contrast, by including animals in our thinking, we extend what we mean by work. What does it mean to say that a cow producing milk for human consumption is "working"? Can we say that wild animals – who may unwittingly serve humans – are working too? Consider the wild birds scattering seeds mentioned above, or perhaps beavers who, in damming rivers, build their own homes while also alleviating flood risks for human neighbours (Welden, 2023). It is important to point out that animal work should not be exclusively defined by human involvement. Many animals care for others and engage in eco-social reproduction far from human oversight (Blattner et al., 2020; Coulter, 2016b). However, much of the research on animals and work focuses on domestic animals because they are more embedded within human social, economic and cultural systems, making their labour more visible and directly tied to human activities.

In particular, horses have long been integral to human work, from their historical roles in agriculture, transportation and warfare to their contemporary involvement in sports. The act of riding opens up distinctive forms of interspecies communication and partnership, allowing humans and horses to physically connect while sharing geographical and emotional experiences (Dashper, 2016; Game, 2001). In the UK and elsewhere, these enduring connections with horses have in turn helped shape "significant and symbolic societal transformations" that have shaped industrial society and impacted both humans and horses (Adelman & Thompson, 2017, p. 3). With their overlapping micro and macro significance, then, horse–human relations are especially useful in developing an interspecies understanding of work.

The nature of work involving horses varies significantly across geographic, sociocultural and economic contexts. In some African or Asian countries, for example, their labour in agriculture, transport and other industries is often critical to sustaining human livelihoods, with both human and animal workers frequently enduring poverty and harsh conditions (Watson et al., 2022). In wealthier countries like the UK, horse–human interactions typically centre on tourism, leisure and sport, where economic stakes are lower and their roles are less about survival (Wadham & Dashper, 2024). Nonetheless, within these industries horses remain vital social and economic contributors. The chapter primarily focuses on the tourism industry, a field that encompasses a wide range of settings in which horse–human labour unfolds and provides rich ground for exploring interspecies work across diverse cultural and economic contexts. As a global industry, tourism also presents a range of ethical dilemmas, shaped by differing cultural attitudes, economic pressures and consumer expectations vis-à-vis human–animal relations (Rickly & Kline, 2021). Finally, the industry is marked by the marginalisation of both human and animal workers, offering a critical lens for examining shared vulnerabilities and interspecies inequalities (Wadham & Dashper, 2024).

Introducing our expert collaborator for this chapter

Professor Kate Dashper is a leading scholar who integrates gender studies and human–animal studies within the fields of tourism, events, leisure and sport. She explores how humans and animals collaborate in work and play and how interspecies relationships emerge through shared activities. Her work demonstrates how multispecies perspectives can deepen our understanding of practices and organisations in these sectors. Her monograph *Human–Animal Relationships in Equestrian Sport and Leisure* (Dashper, 2016) offers one of the first sustained multispecies analyses of leisure and sport. Building on this work, she has co-edited pioneering volumes on multispecies event management (Dashper et al., 2021) and multispecies leisure in Latin America (Monterrubio et al., 2025).

Kate has focused on a range of equine workers, from racehorses (Monterrubio et al., 2023), to those working in trekking centres (Dashper, 2020b) and forest environments (Wadham & Dashper, 2024). While her fieldwork has primarily centred on the UK, she has also studied horse–human work in Iceland (e.g. Helgadóttir & Dashper, 2016) and Mexico (e.g. Monterrubio et al., 2024). Theoretical ideas explored in her human–animal research include emotional labour (Dashper, 2019), interspecies collaboration (Dashper, 2020a) and attentiveness (Dashper, 2017). This chapter draws on much of this work, interweaving it with research by other significant scholars in this field, including Kendra Coulter and Linda Tallberg.

In what follows, we briefly define the concept of work itself, highlighting its anthropocentric limitations. We will then explore why we should include animals within our academic ideas about work, highlighting the practical, ethical and conceptual benefits of doing so. Turning to the research of Kate Dashper and others, we will then explore horses' labour in tourism, leisure and sport and how this helps illuminate and extend our understanding of the very concept of work. Finally, we advocate for future research and teaching practices that embrace animals as co-workers, fostering more inclusive and responsible management practices.

Conceptual foundations: What is work?

"Work" is the conscious activity through which individuals contribute to their own wellbeing and that of others (Timmerman, 2018). It encompasses all purposeful activity that contributes to sustaining or enhancing life, regardless of its economic, formal or physical dimensions. As such, work includes the various forms of paid employment, but also domestic labour, voluntary activities and other unpaid productive and reproductive efforts. With this in mind, it is perhaps unsurprising that the meanings and boundaries of what constitutes work have remained contested and variable.

Moving on from the "two domains" model of work

Historically, dominant discourses about work have revolved around paid labour in the public sphere. The "two domains model," prevalent in 20th-century industrialised nations, dichotomised work and non-work. Paid employment was seen as economically and socially valuable when compared to unpaid activities like domestic labour, caregiving and leisure. This framework reflected and reinforced capitalist values, which see work primarily as a unit of economic exchange (Taylor, 2004). However, feminist scholars in particular have exposed how reducing work to paid employment marginalises not only women but also other groups, including migrants, children, older adults and those unable to engage in formal employment due to illness or disability (Kalleberg, 2009). Glucksmann's (1995) concept of the "Total Social Organisation of Labour" (TSOL) has been especially influential. This points out the diversity and interconnectedness of all forms of work, whether paid/unpaid, formal/informal or public/private.

What are the core criteria that underpin this very broad concept of work? First, it is a *life-sustaining activity*, essential for survival and flourishing (Bourne & Calás, 2013; Williams & Nadin, 2012). Here, both productive work (focused on generating goods and services) and reproductive work (encompassing caregiving, domestic tasks and child-rearing) play critical roles in sustaining individuals and societies. Second, work *creates value across multiple domains*. Non-remunerated efforts such as caregiving and volunteerism, for example, can make a significant contribution to individual, societal and environmental wellbeing. Third, work is *embedded in social relations*. As such, it is shaped by and contributes to relationships within families, organisations and societies (Glucksmann, 1995). Fourth, it is *shaped by power dynamics*. Those who design and control work systems – employers, governments and elites – often have more social and economic influence than those who perform the tasks at hand (Dundon et al., 2020).

Acknowledging work as a culturally specific ideal

As above, dominant "Western" understandings of work remain deeply influenced by capitalism, which conceptualises work as a unit of economic exchange. This perspective is not universal but culturally specific, contrasting with non-capitalist and reciprocal economies where work is closely tied to social roles and communal obligations (Weeks, 2020). An influential group of "post-work" scholars suggests that we need to denaturalise the very notion of work and remove it from the centre of our individual and shared lives (Hoffman et al., 2024).

A key factor in the Western preoccupation with work as the defining feature of life is its anthropocentrism. This is the belief that human beings are the central or most important entities in the world (Kandel et al., 2023). This perspective not only marginalises animals by dismissing their contributions and agency but also perpetuates a broader – potentially catastrophic – mindset that positions humans as separate from and superior to the natural world. Encouraging less human-centred ways of thinking about work is therefore an important step in helping foster more equitable and sustainable ways of living.

Towards an interspecies approach: Work for, with and by animals

There is growing recognition that humans and animals have long collaborated in work, actively shaping labour practices together throughout history and into the present (O'Doherty, 2016; Tallberg & Hamilton, 2022). Coulter (2016a) helpfully distinguishes between work done *for, with* and *by* animals (see also Kelemen et al., 2020). These categories reveal both the persistence of human dominance but also the possibility of reimagining work as something that extends beyond the human species.

Work *for* animals: This centres on the care and welfare of animals, encompassing human jobs, such as vets, zookeepers and shelter workers. These roles often reinforce anthropocentric hierarchies, positioning humans as dominant caregivers within a "natural order." They also reproduce other forms of oppression, including those relating to gender, race and class. For example, even in relatively high-status settings like the vet surgery, such work is often feminised and devalued (Clarke & Knights, 2019; Coulter & Fitzgerald, 2019; Irvine & Vermilya, 2010; Treanor & Marlow, 2021).

This work can be exploitative and harmful to both humans and animals. In horseracing, for example, physical proximity, demanding conditions and the emotional bond that develops between racehorses and stable staff results in their "shared suffering" (Porcher, 2011). For the humans – usually young, female and/or migrant workers – the day starts early, finishes late and the "dirty work" in between is hard and repetitive (Miller, 2013). Focusing on work for animals thus reveals how interspecies care practices are deeply intertwined with broader power dynamics. However, its primary focus on human experiences limits insights into the agency and roles of animals themselves.

Work *with* animals: This refers to labour where humans and animals share tasks to achieve mutual or human-centred goals. That is, they depend on and interact with each other. For example, police dogs assist officers in detecting substances or tracking suspects, contributing sensory skills humans lack (Knight & Sang, 2020; Sanders, 2006). Likewise, horses often contribute both power and knowledge, such as trekking ponies who help "look after" unskilled clients by choosing safer paths (Wadham & Dashper, 2024).

These examples show how both species shape work in meaningful ways, challenging assumptions about human exceptionalism. However, animal workers are often more vulnerable than their human counterparts. For example, trekking horses who are insufficiently calm and friendly may be sold or even euthanised (Dashper, 2020a). Similarly, many police dogs face death rather than retirement: Their skills shift from valued to burdensome, as they are seen as difficult to retrain or rehome (Knight & Sang, 2020). While this research highlights collaboration, a focus on work *with* animals can overlook their independence. Animals often act without human prompting and their efforts and motivations shape relationships in significant ways.

Work *by* animals: Like its human corollary, "animal work" spans a wide range of productive and reproductive activities. It includes physical labour like horses working in tourism or logging (Wadham & Dashper, 2024); service work such as police dogs discussed above; ecological work like bees pollinating crops (Davies & Riach, 2019); symbolic work when zoo-goers project cultural and ecological meanings onto pandas (Guo, 2021); and scientific work like mice or rabbits in experimental laboratories (Davies et al., 2024).

As shown above, animals make physical, emotional and cognitive contributions in various work contexts. Sometimes they work autonomously and even when working with or for people, they might not do so knowingly, as with the bees. While animal work remains underexplored, it merits serious academic attention. Like people, animals actively shape work practices and (often) lack choice in whether they participate (Despret, 2016). Seeing them as skilled workers and organisers in their own right offers new insights to otherwise human concepts (Coulter, 2019; Sayers et al., 2019). We will now explore some of these insights, via reference to the work of horses in particular.

Working with horses: Taking animals seriously as workers

Our starting point, outlined above, is that work is not an exclusively human concept or activity. Other animals, too, are workers. Horses offer a compelling example because of their deep entanglement in the histories of human societies and economies (Birke & Thompson, 2017; Wadham, 2021). Following Kate Dashper's work, we focus here on horses who work in the related industries of tourism, leisure and sport.

Horses work in many ways. This is clear in riding schools, where ponies may work six days a week moving from one lesson to the next, or in racing where strict training regimes apply. Even "leisure" horses – kept privately and mainly engaged in "hacking" around local lanes – are working too. This becomes apparent when we apply the concept of "emotional labour." Like human workers, horses are expected to manage their feelings and display specific emotions in order to fulfil job expectations (Dashper, 2019). Trekking horses, for example, must maintain a calm and friendly demeanour in order to enhance the experience of paying tourists. Even horses kept as "pets" are expected to follow rules that may conflict with their "natural" instincts, such as ignoring fast-moving traffic on the road (Wadham, 2021). Like many people, horses are closely managed. If they misbehave, become ill, injured or too old, they may be removed or even killed (Dashper, 2014, 2019). Their status as commodities – beings who can be bought and sold – adds to their precarious status. Whether in commercial or domestic settings, it is the human partner alone who decides whether the bond continues. If the horse no longer meets expectations, they can be sold or passed on (Birke & Thompson, 2017; Wadham, 2020).

Still, horses contribute more than physical power – they offer knowledge and understanding. For example, they may detect cues that humans miss, preventing accidents or difficult situations. The manager of a beachside trekking centre told us:

> Usually it's fine to ride in the water, in the breakers and that's what tourists want a lot of the time… But sometimes the sand shifts and it's less stable and my horses know that and won't go near the water. They tell us, "no, I'm not stepping on that, it's not safe" and I wouldn't have known that just by looking at the sand.
>
> *(Wadham & Dashper, 2024, p. 11)*

While animal work is often precarious, it can also be highly valued. But even when horses are treated as co-workers, their agency – or their ability to make choices – is tightly constrained. A horse might choose a safer path across shifting sand or forest terrain, but they cannot choose not to work at all. This exposes a fundamental ethical question, namely "what's in it for the horse?" (Birke & Thompson, 2017). Recognising animals as workers demands ethical and practical responsibility: To consider their wellbeing, respond to their needs and ensure they receive appropriate benefits for their labour.

Redefining work as an interspecies endeavour

We have argued that while work is an important concept, it is often understood in anthropocentric terms. Using horses as an example, we have demonstrated that other animals can also be seen as workers. Framing work as an interspecies endeavour is a political act – one that challenges human-centred norms. It expands how we think about work in at least three ways, by questioning our focus on economic productivity, exposing overlapping inequalities and examining the cultural fixation on work as central to life itself. We will consider each in turn below.

First, work is typically framed in terms of economic output. Including animal workers allows for a more nuanced understanding. Horses (and other animals) engage in labour that may not be strictly economic in nature, but is nonetheless deeply valuable to people, organisations and society. Their presence encourages a shift towards work practices that consider their needs and capacities. For example, trekking centres often run to tight schedules and margins, but the rhythms and requirements of the horses come first. Staff teach clients to prioritise the horses – rides don't end until the horse is back in their stable, fed and watered. In some cases, animals work alongside humans. In others, they work independently. Rewilding projects, for example, show how horses' natural behaviours – grazing, trampling and fertilising grasslands – can align seamlessly with human-directed ecological goals (DeSilvey & Bartolini, 2019). This "eco-social reproduction" (Coulter, 2016b) supports both environmental sustainability and human economic interests, often with minimal human–animal interaction. Whether working together or apart, acknowledging animals as workers reminds us that we share and shape this world together.

Second, focusing on animal work helps us see parallels in how both animals and certain groups of people are marginalised – on the basis of gender, race, class or disability. Work involving horses is often poorly paid and carried out by women or people from lower socioeconomic or ethnically minoritised backgrounds. Acknowledging animals as workers makes these intersections visible and prompts us to ask whose labour "counts."

It also destabilises human exceptionalism – the idea that only humans deserve rights and protections. For example, growing concern over what happens to racehorses after retirement reflects an emerging view of horses as athletes with similar needs to their human counterparts. This raises questions about the racing industry's ethical responsibility to support these animals after their careers end. But elite racehorses are only a tiny fraction of the equine population. What about all the "ordinary" horses also nearing retirement? Recognising animal workers brings with it questions of social responsibility.

Third, including animals in our definitions of work challenges the idea that work is a universal or neutral concept. Marxist and critical theorists have long argued that work is a socially and historically specific construct – one that serves those at the top while exploiting others (Dundon et al., 2020). Animals are drawn into these same capitalist logics of production, yet their experiences are often overlooked (Barua, 2016; Welden, 2023).

By thinking beyond the human, we can begin to imagine alternative models of work. For example, many North American Indigenous cultures historically integrated animals into reciprocal relationships rather than exploitative labour systems. Horses were seen as companions and members of the community, not tools or property (Seifert, 2016). These examples show that our work-centred society is not inevitable, but culturally specific – tied to capitalist priorities. This shift in perspective opens space for more inclusive, relational and sustainable understandings of work, value and purpose – across both species and cultures. For managers operating in sectors that depend on animal labour, this reframing has direct consequences. It shifts responsibility from managing resources to coordinating interspecies relations shaped by power, vulnerability and care.

Directions for future research

Throughout this chapter, we've suggested that the concept of animal work invites us to rethink fundamental assumptions about both work and our relationships with other species. This section outlines three key directions for future research: Expanding conceptual frameworks around interspecies work; examining how human–animal work fosters connections with nature; and addressing the ethical and practical challenges of integrating animals into organisational systems.

First, future research could further expand the conceptual boundaries of what constitutes work. Traditional definitions, rooted in anthropocentric and capitalist frameworks, often overlook the diverse ways animals contribute to human and environmental systems. Research that includes animals as active participants could explore the physical, cognitive and emotional dimensions of their labour, across contexts ranging from service and agriculture to ecology and science. Taking animals seriously in this way means attending to how they exhibit agency and decision-making, such as the way horses interpret complex sensory data in policing. This requires interdisciplinary methods that draw from ethology, sociology and organisation studies. Researchers might also consider how animals challenge human-centred notions of skill, productivity and value. Comparative analyses across species and cultures could offer especially valuable insights.

Second, researchers could examine how organisations act as intermediaries that reconnect people with nature through human–animal work. As urbanisation increases, many people lose direct contact with animals. Organisations employing animal workers – especially in tourism or conservation – have a unique role to play in bridging that gap. Farm parks and ecotourism sites, for instance, can foster appreciation for interspecies collaboration and ecological interdependence. Researchers might focus on how such organisations communicate the roles and welfare of animal workers. This could involve examining how storytelling and social media efforts prompt critical reflection on the individuals and species involved. Broader studies could explore how these efforts encourage empathy, ethical responsibility and environmental stewardship.

Finally, future research should address the ethical and practical complexities of integrating animals into human-centred labour systems. Recognising animals as workers helps value their contributions, but it also raises difficult questions about exploitation, welfare and agency. Scholars could explore how businesses balance economic goals with the need for fair and compassionate treatment. This might involve evaluating welfare standards and identifying best practices for embedding ethics into policy. For example, what might "decent work" mean for animals? How can it be measured beyond standard veterinary markers of health? Researchers could also examine the risk of reinforcing exploitation through the commodification of animal labour. Cross-cultural and cross-industry studies could reveal alternative models that prioritise reciprocity and shared benefit over economic gain alone.

Ethical implications

Considering animals as workers is not straightforward. Rather, it raises a number of complex ethical questions.

First, we must navigate tensions between animal wellbeing and human social or cultural practices. In tourism, for example, work can be precarious or even dangerous for both humans and animals (Winchenbach et al., 2021). Recognising this shared vulnerability is essential to improving animal welfare without undermining human livelihoods. Yet developing universal welfare standards is difficult, as ideas about what is acceptable vary across cultural and economic contexts (Blattner et al., 2020). In this light, the concept of a "social licence to operate" may offer a more flexible framework – one that considers local realities while setting ethical baselines for what can reasonably be expected of working animals (Monterrubio et al., 2024).

Second, extending the concept of work to animals risks drawing other species into labour systems defined by inequality and precarity. In other words, recognising animals as workers may simply co-opt them into structures that exploit their human counterparts, thereby legitimising or even deepening those same hierarchies. However, this chapter has argued that we do not live in an ideal world and – whether we acknowledge it or not – animals like horses are already enrolled in these systems. That is, the lives of domesticated animals are already shaped by long-standing dependencies on humans (Barua, 2016; Haraway, 2008). While it's important to reflect on how these relationships came to be, the more urgent task is to ask how do we move forward from here? In the case of interspecies work, our focus should be on making these existing relationships more equitable – improving conditions for animals while also respecting the rights and wellbeing of the humans who work with them.

Practical implications for responsible management

The first and most critical step for industries, organisations and leaders is to acknowledge the presence and contributions of animals within their operations, whether direct or indirect. In some cases, animals' roles are visible, as with police horses or service dogs. In others, they are less apparent, like bees supporting agriculture or mice in biomedical laboratories. Acknowledging animals as not tools but active participants requires a shift in perspective (Sayers et al., 2019). For instance, tourism businesses may depend upon animals like elephants and camels to provide rides to tourists, while overlooking both the ethical implications of their work and the wellbeing of their human handlers. Conversely, in supply chains, animals often support production behind the scenes (e.g. silkworms in textiles) while remaining invisible to end-users. Responsible management demands acknowledging both visible and hidden roles and accepting accountability for all the animals who support business outcomes.

One key step towards this accountability is the development and enforcement of animal welfare standards (see Chapter 12 on animals and company activism). In the case of horses, this might include adequate rest, access to social contact and proper nutrition. Tourism businesses could revise schedules and include recovery periods. Embedding welfare policies into organisational frameworks – aligned where possible with standards for human employees – signals a serious ethical commitment. A holistic understanding of welfare is essential, extending beyond veterinary care to include emotional wellbeing and behavioural indicators. Collaboration with interdisciplinary experts – such as veterinarians, ethologists and social scientists – can ensure that welfare standards reflect animals' lived realities.

It is also important to recognise the connection between human and animal welfare. Stable hands, zookeepers, tourism staff and others working with animals often face low pay, long hours and physical and emotional strain (Winchenbach et al., 2021). Improving working conditions for human employees via fair pay, job security and training can directly enhance animal welfare. A stable worker with time, resources and support is better equipped to care for horses, creating a more positive environment for both. Ethical animal care begins with ethical treatment of human caregivers.

Transparent communication plays a vital role in this process. Public perception matters, especially when animals are involved: Maintaining a business's social licence to operate depends on public trust. Businesses can strengthen this trust by sharing stories, images and updates about their animal workers – highlighting them as valued individuals rather than faceless tools. Some UK police forces, for example, share updates on their horses' workdays and downtime, individualising them and reinforcing public confidence. Sharing welfare benchmarks and openly discussing practices not only builds credibility but also invites accountability.

In a time when ethical concerns increasingly shape public and industry standards, responsible management requires organisations to take animals seriously, not just as symbolic figures or functional assets, but as participants in shared systems of labour and care. In doing so, this chapter repositions animals not as background assets within work systems, but as co-workers whose lives and labour must be central to sustainable and responsible organisational futures.

Reflective activities for inside and outside the classroom

A Revision questions that test knowledge ("what?")

- A1 According to the chapter, how has the traditional concept of work been defined and what are its limitations?
- A2 What is meant by "work for, with and by animals"?
- A3 What is meant by "emotional labour" in the context of horse work?
- A4 What are some of the similarities and differences in the way animals and other specific groups of workers are treated within organisations and society more broadly?

B Reflective questions that build understanding ("so what?")

- B1 Why does recognising animals as workers require us to rethink conventional ideas of agency, consent and welfare?
- B2 In what ways does an interspecies perspective help highlight shared vulnerabilities between human and nonhuman workers?
- B3 How might focusing on animal workers challenge capitalist assumptions about productivity and value?
- B4 Why is it important to consider both visible and invisible forms of animal labour?

C Take action ("do")

- C1 Choose an industry (e.g. tourism, agriculture, research, security) and map the different roles animals play. Identify areas where welfare or recognition could be improved.
- C2 Create a set of practical guidelines or a checklist for an organisation in your chosen industry to improve support for animal workers. Your checklist might include things like adequate rest periods, access to proper nutrition, opportunities for social interaction with other animals, mental stimulation etc. Then, critically assess your approach, i.e. what are the strengths and limitations of this checklist approach for organisations, animals and the people who work with them?
- C3 Create a storyboard or social media campaign that highlights the contributions of equine workers. Focus on how to make horses more

understandable/relatable to the public so that the campaign might help educate a broad audience about their roles and welfare needs. What are the strengths/limitations of this "humanising" approach to animal workers?

- C4 Write/present a mini literature review exploring how different cultures understand and value animal labour.

D Ideas to discuss with colleagues, friends and family ("share")

- D1 Have you ever interacted with horses, such as through riding lessons, tourism activities, or attending equestrian events? Would you consider what these horses were doing to be "work"? Why (not)?
- D2 Have you seen any stories or videos on platforms like Instagram or TikTok about horses in tourism or sports? How did these narratives influence your views on the treatment of horses as workers or co-workers in these industries?
- D3 Discuss with friends or family how your respective childhood and cultural experiences shape your (perhaps different) assumptions about animals and the work they can/should do.
- D4 What would fair compensation look like for animal workers? What kind of care or support should they receive when they retire or can no longer work? In what ways do these questions compare to how we treat human workers?

Key ideas

Anthropocentrism

A belief system that places humans at the centre of importance, often ignoring or devaluing other species. It shapes dominant ideas about work by sidelining nonhuman contributions and reinforcing human-centric narratives (Haraway, 2008; Kandel et al., 2023).

Anthropomorphism

Attributing human thoughts, emotions or motivations to nonhuman beings. While it can promote empathy, it may also distort how we understand animals by projecting human perspectives onto them instead of recognising their own distinct ways of being (Serpell, 2003).

Marginalisation

The process by which individuals or groups are pushed to the edges of society and denied full participation. In interspecies work, both animals and precarious human workers often face intersecting forms of exclusion that reduce their visibility, voice and value (Coulter, 2016a; Hribal, 2003; Tallberg & Hamilton, 2022).

Agency

The ability to make choices and influence outcomes. Recognising animals' agency challenges human-centred assumptions by acknowledging that animals can actively shape work environments, respond to situations and express preferences in ways that reflect their own needs and capacities. However, in many work contexts, this agency is significantly constrained by human control, economic pressures and institutional structures that limit animals' ability to act freely or opt out of labour altogether (Birke & Thompson, 2017; Despret, 2016).

Social licence to operate

The informal trust and approval from the public or local communities that allows an organisation to continue its work. When animals are involved, this trust often depends on transparency, ethical treatment and efforts to address concerns about welfare and exploitation (Monterrubio et al., 2024).

References

Adelman, M., & Thompson, K. (Eds). (2017). *Equestrian cultures in global and local contexts*. Springer. doi:10.1007/978-3-319-55886-8.

Barua, M. (2016). Lively commodities and encounter value. *Environment and Planning D: Society and Space*, 34(4), 725–744.

Birke, L., & Thompson, K. (2017). *(Un)Stable relations: Horses, humans and social agency*. Routledge.

Blattner, C. E., Coulter, K., & Kymlicka, W. (Eds). (2020). *Animal labour: A new frontier of interspecies justice?* Oxford University Press. doi:10.1093/oso/9780198846192.001.0001.

Bourne, K. A., & Calás, M. B. (2013). Becoming "real" entrepreneurs: Women and the gendered normalization of "work." *Gender, Work & Organization*, 20(4), 425–438.

Clarke, C. A., & Knights, D. (2019). Who's a good boy then? Anthropocentric masculinities in veterinary practice. *Gender, Work & Organization*, 26(6), 886–902. doi:10.1111/gwao.12244.

Coulter, K. (2016a). *Animals, work and the promise of interspecies solidarity*. Palgrave Macmillan.

Coulter, K. (2016b). Beyond human to humane: A multispecies analysis of care work, its repression and its potential. *Studies in Social Justice*, 10(2), 199–219. doi:10.26522/ssj.v10i2.1350.

Coulter, K. (2019). Toward humane jobs and work-lives for animals. In C. Blattner, K. Coulter, & W. Kymlicka (Eds), *Animal labour: A new frontier of interspecies justice?* Oxford University Press (pp. 29–47). doi:10.1093/oso/9780198846192.003.0002.

Coulter, K., & Fitzgerald, A. (2019). The compounding feminization of animal cruelty investigation work and its multispecies implications. *Gender, Work & Organization*, 26, 288–302. doi:10.1111/gwao.12230.

Dashper, K. (2014). Tools of the trade or part of the family? Horses in competitive equestrian sport. *Society & Animals*, 22(4), 352–371. doi:10.1163/15685306-12341343.

Dashper, K. (2016). *Human–animal relationships in equestrian sport and leisure*. Routledge. doi:10.4324/9781315678139.

Dashper, K. (2017). Listening to horses: Developing attentive interspecies relationships through sport and leisure. *Society & Animals*, 25(5), 475–493. doi:10.1163/15685306-12341426.

Dashper, K. (2019). Moving beyond anthropocentrism in leisure research: Multispecies perspectives. *Annals of Leisure Research*, 22(2), 133–139. doi:10.1080/11745398.2018.1478738.

Dashper, K. (2020a). More-than-human emotions: Multispecies emotional labour in the tourism industry. *Gender, Work & Organization*, 27(1), 24–40. doi:10.1111/gwao.12344.

Dashper, K. (2020b). Holidays with my horse: Human–horse relationships and multispecies tourism experiences. *Tourism Management Perspectives*, 34. doi:10.1016/j.tmp.2020.100678.

Dashper, K., Helgadóttir, G., & Sigurðardóttir, I. (Eds). (2021). *Humans, horses and events management*. CABI. doi:10.1079/9781789242751.0000.

Davies, G., Greenhough, B., Hobson-West, P., Kirk, R. G., Palmer, A., & Roe, E. (2024). *Researching animal research*. Manchester University Press. doi:10.7765/9781526165770.

Davies, O., & Riach, K. (2019). From manstream measuring to multispecies sustainability? A gendered reading of bee-ing sustainable. *Gender, Work & Organization*, 26(3), 246–266. doi:10.1111/gwao.12245.

DeSilvey, C., & Bartolini, N. (2019). Where horses run free? Autonomy, temporality and rewilding in the Côa Valley, Portugal. *Transactions of the Institute of British Geographers*, 44 (2), 313–327. doi:10.1111/tran.12251.

Despret, V. (2016). *What would animals say if we asked the right questions?* University of Minnesota Press. doi:10.5749/minnesota/9780816692378.001.0001.

Donaldson, S., & Kymlicka, W. (2011). *Zoopolis: A political theory of animal rights*. Oxford University Press.

Dundon, T., Lucio, M. M., Hughes, E., Howcroft, D., Keizer, A., & Walden, R. (2020). *Power, politics and influence at work*. Manchester University Press. doi:10.7765/9781526157119.

Game, A. (2001). Riding: Embodying the centaur. *Body & Society*, 7(4), 1–12. doi:10.1177/1357034X01007004001.

Glucksmann, M. (1995). Why "work"? Gender and the "total social organization of labour." *Gender, Work & Organization*, 2(2), 63–75. doi:10.1111/j.1468-0432.1995.tb00028.x.

Guo, Y. (2021). "Cute, but get up and work!" The biophilia hypothesis in tourists' linguistic interactions with pandas. In J. M. Rickly & C. Kline (Eds), *Exploring non-human work in tourism* (pp. 225–242). De Gruyter.

Haraway, D. J. (2008). *When species meet*. University of Minnesota Press.

Helgadóttir, G., & Dashper, K. (2016). "Dear international guests and friends of the Icelandic horse:" Experience, meaning and belonging at a niche sporting event.

Scandinavian Journal of Hospitality and Tourism, 16(4), 422–441. doi:10.1080/15022250.2015.1112303.
Hoffman, M., Pantazidou, M., & Smith, T. (2024). Three critiques of work: The radical roots of degrowth. In L. Eastwood & K. Heron (Eds), *De Gruyter handbook of degrowth* (pp. 55–74). De Gruyter. doi:10.1515/9783110778359-006.
Hribal, J. (2003). "Animals are part of the working class": A challenge to labour history. *Labor History*, 44(4), 435–453. doi:10.1080/0023656032000170069.
Irvine, L., & Vermilya, J. R. (2010). Gender work in a feminized profession: The case of veterinary medicine. *Gender & Society*, 24(1), 56–82. doi:10.1177/0891243209355978.
Kalleberg, A. L. (2009). Precarious work, insecure workers: Employment relations in transition. *American Sociological Review*, 74(1), 1–22. doi:10.1177/000312240907400101.
Kandel, I. J., Dlouhy, K., & Schmitt, A. (2023). Animal roles in organisations: A framework for exploring organizational human–animal relations. *Organization*, 32(4), 508–526. doi:10.1177/13505084231217079.
Kelemen, T. K., Matthews, S. H., Wan, M., & Zhang, Y. (2020). The secret life of pets: The intersection of animals and organisational life. *Journal of Organisational Behaviour*, 41(7), 694–697. doi:10.1002/job.2465.
Knight, C., & Sang, K. (2020). "At home, he's a pet, at work he's a colleague and my right arm:" Police dogs and the emerging posthumanist agenda. *Culture and Organization*, 26(5–6), 355–371. doi:10.1080/14759551.2019.1622544.
Miller, J. (2013). Racing bodies. In C. Wolkowitz, R. Cohen, T. Sanders, & K. Hardy (Eds), *Body/sex/work: Intimate, embodied and sexualised labour* (pp. 193–206). Palgrave. doi:10.1007/978-1-137-02191-5_12.
Monterrubio, C., Dashper, K., & Hernández-Espinosa, R. (2023). Horseracing as gendered leisure events: Doing and redoing masculinities in Mexico. *Leisure Studies*, 42(4), 553–567. doi:10.1080/02614367.2022.2115115.
Monterrubio, C., Dashper, K., & Wadham, H. (2024). Contested intangible heritage: Equestrian sport and animal welfare in Mexico. *International Journal of Heritage Studies*, 30(3), 332–348. doi:10.1080/13527258.2023.2294761.
Monterrubio, C., López-Medellín, X., Dashper, K., & Wadham, H. (Eds). (2025). *Human–animal relations in tourism, leisure and development: Perspectives from Latin America*. CABI. doi:10.1079/9781800626829.0000.
O'Doherty, D. (2016). Feline politics in organisation: The nine lives of Olly the cat. *Organization*, 23(3), 407–433. doi:10.1177/1350508416629450.
Porcher, J. (2011). The relationship between workers and animals in the pork industry: A shared suffering. *Journal of Agricultural and Environmental Ethics*, 24(1), 3–17. doi:10.1007/s10806-010-9232-z.
Rickly, J. M., & Kline, C. S. (Eds). (2021). *Exploring non-human work in tourism: Understanding interspecies relationships*. Routledge.
Sanders, C. R. (2006). The dog you deserve: Ambivalence in the K-9 officer/patrol dog relationship. *Journal of Contemporary Ethnography*, 35(2), 148–172. doi:10.1177/0891241605283456.
Sayers, J., Hamilton, L., & Sang, K. (2019). Organizing animals: Species, gender and power at work. *Gender, Work & Organization*, 26(3). doi:10.1111/gwao.12277.
Seifert, A. R. (2016). An introduction to contemporary Native American horse culture: Notes from the Northwest Plateau. In D. Davis & A. Maurstad (Eds), *The meaning of horses' biosocial encounters* (pp. 147–163). Routledge.

Serpell, J. A. (2003). Anthropomorphism and anthropomorphic selection: Beyond the "cute response." *Society & Animals*, 11(1), 83–100. doi:10.1163/156853003321618864.

Tallberg, L., & Hamilton, L. (Eds). (2022). *The Oxford handbook of animal organization studies*. Oxford University Press. doi:10.1093/oxfordhb/9780192848185.001.0001.

Tallberg, L., Välikangas, L., & Hamilton, L. (2022). Animal activism in the business school: Using fierce compassion for teaching critical and positive perspectives. *Management Learning*, 53(1), 55–75. doi:10.1177/13505076211044612.

Taylor, S. (2004). The right not to work: Power and disability. *Monthly Review*, 55 (10), 36–44. doi:10.14452/MR-055-10-2004-03_2.

Timmerman, C. (2018). Contributive justice: An exploration of a wider provision of meaningful work. *Social Justice Research*, 31, 85–111.

Treanor, L., & Marlow, S. (2021). Paws for thought? Analysing how prevailing masculinities constrain career progression for UK women veterinary surgeons. *Human Relations*, 74(1), 105–130. doi:10.1177/0018726719846554.

Wadham, H. (2020). Horse matters: Re-examining sustainability through human–domestic animal relationships. *Sociologia Ruralis*, 60, 530–550. doi:10.1111/soru.12293.

Wadham, H. (2021). Relations of power and nonhuman agency: Critical theory, Clever Hans and other stories of horses and humans. *Sociological Perspectives*, 64 (1), 109–126. doi:10.1177/0731121420921897.

Wadham, H., & Dashper, K. (2024). The slow road to sustainability: An interspecies perspective on decent work. *Journal of Sustainable Tourism*, 33(9), 1816–1831. doi:10.1080/09669582.2024.2374361.

Watson, T., Kubasiewicz, L. M., Nye, C., Thapa, S., Norris, S. L., Chamberlain, N., & Burden, F. A. (2022). "Not all who wander are lost:" The life transitions and associated welfare of pack mules walking the trails in the Mountainous Gorkha region, Nepal. *Animals*, 12(22), 3152. doi:10.3390/ani12223152.

Weeks, K. (2020). *The problem with work: Feminism, Marxism, antiwork politics and postwork imaginaries*. Duke University Press. doi:10.1515/9780822394723.

Welden, E. A. (2023). Conceptualising multispecies collaboration: Work, animal labour and nature-based solutions. *Transactions of the Institute of British Geographers*, 48(3), 541–555. doi:10.1111/tran.12593.

Williams, C. C., & Nadin, S. (2012). Work beyond employment: Representations of informal economic activities. *Work, Employment and Society*, 26(2), 353–364. doi:10.1177/0950017012437006.

Winchenbach, A., Hanna, P., & Miller, G. (2021). Rethinking decent work: The value of dignity in tourism employment. *Journal of Sustainable Tourism*, 29(1), 1–19. .

6

ANIMALS AND SOCIAL MEDIA

Introduction

The internet, social media, mobile applications and other forms of digital communication technologies have become deeply embedded in the daily lives of billions of people worldwide (Dwivedi et al., 2021). As of late 2025, an estimated 5.5–6 billion individuals are active internet users, of whom approximately 5.5 billion use social media (Statista, 2025). Together, these figures account for roughly 67–75% of the global population. Social media, in particular, has become a central component of contemporary social interaction, shaping how individuals connect, share information and construct their personal and collective identities. At the same time, companies increasingly rely on social media platforms for marketing communication, where the content created and shared plays a vital role in the success of their marketing efforts (Dwivedi et al., 2021). Social media has also been conceptualised as an arena for everyday conversation – spaces where participants informally discuss and share opinions on current issues, thereby contributing to deliberation about what constitutes (ir-)responsible business conduct (Lundgaard & Etter, 2023). In this regard, social media not only influences business practices and education but also civic engagement and public discourse, underscoring its role in shaping modern society.

While social media may encompass a plurality of voices (Schultz et al., 2013) and reflect heterogeneity in norms, values, expectations and concerns (Etter et al., 2018), it has largely been conceptualised as a space for human interaction and communication. This raises the question of what role animals play within social media environments. Drawing on the experience of expert collaborator Meghan Shaw, this chapter explores social media as an interspecies space and considers the ethical implications of how animals are represented and made visible online.

DOI: 10.4324/9781003585541-9

A growing number of cases illustrate how animals have challenged the anthropocentric assumptions underpinning social media. The case of Cecil the Lion is one of the most striking examples of how social media can mobilise global attention towards animal welfare and ethics (see Mkono, 2018; Mkono & Holder, 2019). Cecil was a well-known, radio-collared lion living in Hwange National Park, Zimbabwe and part of a long-term Oxford University research project. In 2015, he was killed by an American trophy hunter, which triggered an outpouring of outrage across social media platforms (Makuyana, 2018). The case of Naruto, a macaque, who in 2011 took a series of photographs using a camera left unattended by wildlife photographer David Slater in Indonesia, raised intriguing questions about animal agency. One of these photos – a self-portrait of Naruto – went viral online, sparking widespread fascination with the idea of an animal "creating" social media content and having authorship rights.

Both examples highlight the roles of human–animal relationships and animal agency in the digital age (Mkono, 2018; Mkono & Holder, 2019). The killing of Cecil the Lion provoked a powerful collective emotional reaction across social media platforms, shaping public debate, raising conservation awareness and prompting corporate responses. For instance, several airlines reviewed their policies on transporting hunting trophies following public attention on social media (Makuyana, 2018). The case of Naruto prompted discussions about animal agency, authorship and the anthropomorphisation of animals in media. From a social media perspective, the cases of Cecil and Naruto show how animals can be central actors in digital culture, influencing online engagement, viral trends and even debates about the legal and ethical responsibilities towards nonhuman participants.

Despite the strong presence of animals and other species on social media, this digital environment remains largely human-centred in terms of content creation, privacy concerns, ethical considerations and perceived agency. The impacts of social media on animals and other species are often overlooked, reinforcing human exceptionalism in spaces that increasingly shape societal norms and values. By failing to recognise animals as central actors in social media discussions, we not only risk excluding them from moral and ethical consideration but also miss an important opportunity to leverage these platforms to promote human behaviours that contribute to the wellbeing of different species. Moreover, highlighting animals as active participants rather than passive objects or mere content opens the door to more responsible digital practices and ethical frameworks. Indeed, social media could serve as a space for critical reflexivity, providing an opportunity to examine our ethical responsibilities towards other species and to foster more equitable and sustainable interspecies relationships. This chapter contributes to the book's broader project of developing an interspecies approach to responsible management by examining how digital platforms shape human–animal relations and corporate responsibility.

Our expert collaborators and the case

This chapter is based on an interview with Meghan Shaw, who has a background in zoology, marketing and behavioural sciences. She worked as a zookeeper with big cats at a South African zoo, where she observed that visitors wanted to take action for wildlife but often didn't know how. This experience deepened her interest in communication, conservation and biodiversity. Since then, Meghan has worked on a range of projects, including understanding public opinions on biodiversity and policy, addressing various human–wildlife conflicts and exploring ways to encourage people to take everyday actions to protect nature. In her doctoral dissertation, she explored how social media imagery influences people's perceptions and behaviours towards native wildlife, including the negative effects of wildlife selfies and how social media can be used to improve the situation of endangered and overlooked species. In addition to her academic work, she served as president of the Conservation Marketing and Engagement Working Group for the Society for Conservation Biology and has volunteered as an Educator for various wildlife charities. Through her research and conservation initiatives, she has influenced zoo guidelines and international policies on wildlife tourism practice.

This chapter builds on Meghan's work as both a scholar and a wildlife conservationist, examining the role of social media images in promoting tolerance and positive attitudes towards diverse species. Through her research, she demonstrates that photos showing humans in close proximity to wildlife can elicit a range of problematic audience perceptions – for example, that an animal is less endangered or would make a suitable pet – which, in turn, can fuel the illegal wildlife trade. She also found that images featuring humans and animals do not receive more likes or comments than photos of animals alone. This suggests that avoiding human–wildlife content and the potential harms associated with it does not negatively affect social media engagement. On the other side, her findings show that social media can be a powerful tool for encouraging pro-conservation behaviour when content evokes strong, positive emotions. With so much wildlife conservation messaging focused on "doom and gloom," her work suggests that an approach rooted in conservation optimism may be far more effective. Using wildlife conservation as a case study, this chapter delves into Meghan's work, demonstrating how her experiences and research contribute to a deeper understanding of social media from an interspecies perspective. Finally, the chapter calls for future research that adopts a non-anthropocentric perspective, treating social media as a space that can help recognise the interests and agency of nonhuman animals. For managers and organisations, this means that social media strategy is not neutral. Decisions about imagery, messaging and engagement can directly influence animal welfare, conservation outcomes and public norms.

Conceptual foundations: What is social media?

The term "social media" was first introduced in 1994 within the Tokyo-based online environment Matisse, coinciding with the emergence of the first commercial internet platforms that would later define this field (Aichner et al., 2021). In academia, social media has typically been used as an umbrella term encompassing a wide range of online platforms, including blogs, forums, microblogs, photo- and video-sharing sites, product review platforms, social bookmarking services, social gaming environments and virtual worlds (Aichner & Jacob, 2015). Although there is an extensive body of research on social media, there is no universally accepted definition of the concept (Aichner et al., 2021; Carr & Hayes, 2015; Effing et al., 2011). In general, most definitions converge on the idea that social media refers to digital technologies that emphasise user-generated content, networking and interaction (e.g. Aichner et al., 2021; Kaplan & Haenlein, 2010). For example, Carr and Hayes (2015, p. 50) define social media as "internet-based channels that allow users to opportunistically interact and selectively self-present, either in real time or asynchronously, with both broad and narrow audiences who derive value from user-generated content and the perception of interaction with others."

Carr and Hayes (2015) identify five key elements that define social media. First, social media refers to online tools and platforms that operate via the broader internet infrastructure, enabling connectivity across geographic and temporal boundaries. Second, social media channels are persistently available online spaces where content can be created, shared and consumed, regardless of whether users are actively participating at any given moment. Third, a central feature of social media is that users perceive themselves as interacting with others, even if the creation and sharing of content does not always meet traditional criteria for direct interaction. Fourth, the value of social media largely derives from user-generated content and interactions, rather than from content produced by the organisations or platforms that host the space. Finally, social media functions as a platform for mass-personal communication, allowing users to create content that simultaneously engages audiences at both the mass and interpersonal levels, blending personal and public communication in novel ways. Reaching a consensus on the definition of social media will become increasingly challenging as emerging technologies, such as artificial intelligence, introduce new elements and continue to transform its functions.

Social media and responsible business

Social media has created new opportunities for companies to communicate and interact with customers, employees and a wide range of stakeholders (Aichner et al., 2021). In this regard, social media plays a key role in sustainability and corporate social responsibility communication by enabling stakeholder engagement, thereby enhancing the transparency, credibility and legitimacy of business organisations. For example, responsible management scholars have examined stakeholder-generated social media content to assess organisational communication strategies (e.g. Castelló et al., 2016; Colleoni, 2013) and explored how different stakeholders use online platforms to express concerns, opinions and expectations towards organisations (e.g. Etter & Vestergaard, 2015; Hunter et al., 2008). These studies have provided valuable insights into how the notion of (ir-)responsible business is deliberated and constructed within an increasingly transnational, fragmented, pluralistic and complex online space (Lundgaard & Etter, 2023). Moreover, this body of literature shows how social media gives voice to previously disenfranchised actors while also creating more direct and less hierarchical channels through which they can engage with organisations and negotiate social responsibility agendas (e.g. Castelló et al., 2016; Colleoni, 2013; Lundgaard & Etter, 2023).

While the idea of social media as an arena for ethical deliberation – where marginalised stakeholders are given a voice – represents an interesting development with the potential to create more democratic communication spaces, the discussion has remained largely anthropocentric, paying limited attention to the role of animals in relation to social media and responsible business. This is not surprising, given that social media platforms are fundamentally designed for human communication and interaction. This raises an important question: how can animals be considered part of such an inherently human-centric concept? The role of animals in shaping responsible business discussions on social media has been explored, for example, within tourism studies (e.g. Kredens & Vogt, 2023; Mkono, 2018; Mkono & Hughes, 2024). For example, Mkono and Holder (2019) conceptualise social media as digital spaces for collective moral reflexivity concerning human–animal relations. The connection between responsibility and animals in social media has been examined in the field of wildlife conservation, where scholars emphasise the potential of these platforms to encourage pro-conservation behaviours and drive policy changes (e.g. Bergman et al., 2022). In the following section, we examine wildlife conservation as a lens for developing and illustrating the concept of an interspecies perspective on social media.

An interspecies perspective on social media

Drawing on the case of wildlife conservation and insights from the interview with Meghan Shaw, the following discussion outlines three key principles. Together, these highlight how social media can shape human–animal relations, influence pro-animal behaviour and support more inclusive conservation marketing outcomes.

Creating space for better understanding and connection with other species

Social media offers opportunities to learn about species that many people may otherwise never encounter, particularly those that are geographically distant, elusive, or rarely represented in mainstream media. Through text, images and videos, these platforms enable users to connect with animals while providing insights into their lives, habitats and species-specific behaviours. Unlike traditional media formats, social media facilitates repeated, informal and emotionally engaging encounters with wildlife, which can gradually shape attitudes, perceptions and values. Research suggests that increased exposure to and familiarity with a species makes people more likely to value it for characteristics beyond physical appearance, such as ecological function, behavioural complexity or evolutionary uniqueness (Lundberg et al., 2019). As familiarity grows, species are less likely to be judged solely on aesthetic appeal and more likely to be recognised as integral components of broader ecosystems. In this way, social media can play a critical role in expanding public understanding of biodiversity and conservation, while encouraging more inclusive attitudes towards species that are often overlooked or undervalued.

This process aligns with theories of repeated exposure and message reinforcement, which suggest that repeated encounters with an issue or entity increase familiarity, engagement and the likelihood of support over time (Shaw et al., 2022a, 2022b). As Shaw et al. (2022a, 2022b) point out, the deliberate efforts to promote underrepresented species can significantly improve public support, including measurable increases in willingness to donate to conservation causes by approximately a quarter in some cases. All this highlights the potential of social media to reshape human–species relationships through sustained, meaningful exposure rather than isolated information campaigns.

Building on this, Shaw et al. (2024) argue that aesthetic judgements play a significant role in shaping human empathy towards animals and, in turn, influence which species people care about and support. Social media can therefore function as a powerful tool for fostering appreciation and respect for species that are less popular, less visible, or socially accepted (Shaw et al., 2022b). Through storytelling practices, animals can be framed as relatable subjects rather than distant or abstract biological entities, enabling audiences to form emotional connections with species they might otherwise overlook. Narratives that draw on familiar human experiences, such as struggle, resilience, care or adaptation, can generate empathy even for species that are not traditionally perceived as attractive or charismatic. This may involve highlighting distinctive traits, such as an unusual ability or colouration, or emphasising the species' essential ecological role within its ecosystem. By shifting attention away from purely aesthetic appeal and towards relational, functional or moral value, social media storytelling can challenge narrow definitions of which species are considered worthy of concern.

The need to encourage people to relate to different species becomes evident when examining patterns in conservation policy and practice. Governments and conservation organisations have historically prioritised the protection of species that are well known, charismatic, or widely appreciated by the public, often at the expense of less familiar or less aesthetically appealing species (e.g. Donaldson et al., 2016; Troudet et al., 2017). In this context, social media has the potential to promote greater inclusivity by increasing the visibility of species that are disadvantaged in terms of public recognition and emotional appeal. Resources such as donations, media coverage, research attention and grant funding are disproportionately directed towards species that people already recognise and like, while many ecologically important but less familiar species receive limited support. By enhancing visibility and fostering emotional engagement, an interspecies perspective on social media can help rebalance these priorities, supporting more equitable conservation outcomes that extend beyond charismatic animals to encompass a broader range of species and ecosystems. This is a way to foster greater equality and inclusivity among species.

Influencing human behaviour that contributes positively to different species

Social media has become one of the most powerful tools for disseminating messages that can influence human behaviour in ways that positively impact animals and biodiversity. Unlike traditional one-way advertising or communication channels, social media enables two-way interaction, allowing organisations to receive immediate feedback, foster dialogue and build communities around shared values and conservation goals. These interactive affordances make social media particularly well-suited to influencing attitudes, norms and behaviours related to conservation and the protection of different species. Achieving meaningful and sustained behaviour change requires an interdisciplinary approach that integrates insights from both the behavioural sciences and the social sciences. Behavioural science is the organised study of human and animal behaviour through controlled, systematic structure, while social science refers to the study of society, culture and human relationships (see Adhikari, 2016). These disciplines intersect in the field of social marketing, which applies marketing principles to address social problems, design targeted messages and evaluate their effectiveness in influencing behaviour for social good (see Smith et al., 2020).

Conservation marketing builds on these foundations but applies them specifically to the context of biodiversity conservation (Smith et al., 2020). As articulated by the Steering Committee as part of the charter for the (then) provisional Conservation Marketing and Engagement Working Group within the Society for Conservation Biology, Conservation Marketing refers to the ethical application of marketing strategies, concepts and techniques to influence attitudes, perceptions and behaviours of individuals and ultimately societies, with the objective of advancing conservation goals (Wright et al., 2015, p. 46). Conservation marketing goes beyond raising awareness; it prioritises behaviour change by identifying specific target behaviours, understanding audience barriers and motivations and designing interventions that align conservation goals with audience values and everyday practices.

Within this framework, social media plays a central role as both a communication channel and a social environment in which the meaning of conservation is constructed and negotiated. A growing body of research has explored the effectiveness of social media in engaging audiences and promoting pro-conservation behaviours and a positive attitude towards other species (e.g. Ballejo et al., 2021; Kidd et al., 2018; Shaw et al., 2022b). Much of this research has focused on conservation messaging strategies, including how different narratives, emotional appeals, visual representations of species and framing techniques influence audience engagement, trust and support. Findings suggest that storytelling approaches that humanise conservation issues, highlight relational connections between people and species and foreground shared values are more likely to foster an organisation's social licence to operate and encourage positive behaviours towards other species. Overall, an interspecies perspective on social media emphasises influencing human behaviour not merely through information provision, but through carefully designed, ethically grounded strategies that acknowledge the complex psychological and social processes shaping how people relate to animals and act on their behalf.

Acknowledging the unintended consequences of social media

When adopting an interspecies perspective on social media within the context of responsible business, it is crucial to acknowledge that animal-related content can also generate outcomes that are detrimental to both humans and nonhuman species. The visibility of a species on social media does not inherently translate into effective conservation or positive welfare outcomes. In fact, heightened attention can sometimes create misleading perceptions of abundance, normalise exploitative practices and encourage behaviours that compromise animal wellbeing and human safety. Therefore, the mere popularity of animal imagery or narratives online should not be automatically interpreted as indicating positive consequences for animal wellbeing. For example, the notion of paradoxical extinction of charismatic wildlife illustrates well how public perception can inadvertently undermine conservation efforts (Courchamp et al., 2018). Highly recognisable species such as tigers, polar bears and pandas are often assumed to be less endangered than they truly are. This misconception likely stems from their frequent representation in social media and conservation branding – such as the iconic panda logo of the World Wide Fund for Nature – which creates the illusion that there are many of that species in the wild (Courchamp et al., 2018). Repeated exposure to these images can unintentionally give the impression that the species are not of conservation concern (e.g. threatened, endangered or critically endangered), leading to a decline in public urgency and support for their protection, despite their actual vulnerability.

The presence of different species on social media can also contribute to the commodification of wildlife and fuel the illegal pet trade (Collard, 2020; Shaw et al., 2022a; Siriwat & Nijman, 2018). Over the last few years, social media has contributed to an increasing demand for exotic pets, which constitutes one of the most significant threats to global biodiversity (Siriwat & Nijman, 2018). Visual content depicting individuals interacting with wild animals often conveys the misleading impression that these species are suitable as pets. This normalisation encourages prospective buyers to seek out exotic animals, while social media platforms themselves often serve as informal marketplaces where transactions occur between sellers and buyers (Siriwat & Nijman, 2018). For example, Australia has reported some of the highest rates of illegal reptile exports, a trend strongly correlated with social media-driven demand (Chekunov et al., 2024). Similarly, wildlife tourism is shaped by online portrayals of close encounters with animals – such as tourists posing with tigers or riding elephants – which perpetuates the notion that such practices are acceptable (van der Meer et al., 2019). These representations rarely disclose the exploitative conditions behind the scenes, including the use of drugs or coercive training methods to subdue animals (Kredens & Vogt, 2023).

Influencer culture further amplifies these dynamics by commodifying wildlife as props for engagement. A recent case involved an American influencer filmed seizing a wild baby wombat while its distressed mother pursued her (Bucci, 2025). Although this incident provoked widespread condemnation in Australia, public reactions often vary according to species. If the animal had been a snake rather than a wombat, the backlash might have been negligible, revealing a form of speciesism wherein aesthetically appealing animals elicit greater empathy and protection than less charismatic species. The repercussions of social media-driven wildlife interactions extend beyond animal welfare to human safety. In pursuit of striking photographs, individuals have been documented leaving vehicles during safaris or entering zoo enclosures, sometimes with fatal consequences (Daly, 2019). One notable incident involved a man who drowned while attempting to take selfies and videos with a walrus at a zoo in China (Williams, 2016). Such behaviours not only jeopardise human lives but also impose significant stress and harm on the animals involved. These trends underscore the complex and often detrimental role of social media in shaping human–wildlife relationships, requiring more rigorous ethical guidelines, stronger monitoring and public awareness initiatives to mitigate potential negative impacts (see Burns, 2015; Mkono & Hughes, 2024).

Directions for future research

As we have seen throughout this chapter, although different species are present on social media, much remains to be explored regarding the roles animals play as social media actors. With the rapid development of new technologies such as artificial intelligence (AI) and their integration into social media platforms, it will be increasingly important to explore how these technologies shape the types of animal-related content circulating online. AI can be used to filter content or identify the target audiences for specific messages – for example, AI-driven targeting makes it easier to create a wildlife conservation campaign aimed at women aged 25–30. On the other hand, AI may also play a role in flagging or removing problematic animal content. Images of people holding wildlife or interacting with endangered species could be detected by AI and removed immediately, flagged with warnings, or even trigger a notification to the uploader indicating that the content may be inappropriate. Future research could therefore examine how AI might be used to evaluate animal-related content on social media, as well as who should be responsible for this type of monitoring and surveillance. Indeed, there is currently no law specifying which animal-related content is acceptable on social media. Developing clearer practices and ethical guidelines will require research that regards animals as subjects of moral concern. Until now, most discussions and ethical deliberations about social media have focused primarily on its impact on humans, leaving the nonhuman dimension largely unexplored.

In the future, scholars could also critically examine how notions of consent in online spaces might be extended beyond the human, particularly in relation to photographing and filming wildlife, as well as using animals in digital storytelling. While animals cannot provide consent in the conventional sense, researchers could explore alternative models that account for animal agency, welfare and the contexts in which images or videos are produced. This may include identifying behavioural indicators of discomfort, assessing the conditions under which recordings are made, or evaluating the power dynamics between humans and animals in content that violates animal privacy. Scholars could also investigate how ethical responsibilities should be allocated among content creators, platform operators and audiences and consider whether guidelines or regulatory frameworks are needed to protect animals from exploitation or misrepresentation online. This line of inquiry would contribute to broader debates on multispecies ethics and the moral obligations humans hold in digital environments.

Another avenue for future research is the development of multidisciplinary theoretical frameworks to understand how social media contributes to the representation of animals in society. By integrating insights from multiple disciplines, scholars can better interpret a rapidly evolving digital landscape in which animals and other species increasingly appear not only as symbols or content but as mediated actors whose online presence has social, political and ecological consequences. There is also a significant opportunity for methodological innovation. For example, combining netnography (Kozinets, 2015) with multispecies ethnography (Hamilton & Taylor, 2017; Huopalainen et al., 2025) could provide deeper insight into how human–animal interactions and relationships unfold on social media. A multispecies netnographic approach would allow researchers to examine how digital representations shape practices of care, conservation and even exploitation. This line of inquiry could elucidate how animals become active participants, affective agents or even influencers within online environments and how humans navigate these entangled relationships across both digital and physical spaces.

Ethical implications

While an interspecies perspective on social media helps recognise animals and other species as co-creators and actors in online spaces, it also raises important ethical questions that need to be considered. In this chapter, we highlight three key concerns. One of these concerns involves the privacy and datafication of animals, who are increasingly treated as data subjects through tagging, GPS tracking, live cameras and biometric monitoring in social media contexts. This raises questions about data ownership and control: Who owns the data, how is it used or shared and who benefits economically if it is commercialised? Moreover, datafication may expose animals to new risks – such as poaching through geolocation – or unintentionally reinforce their economic value over their intrinsic worth (see Bergman et al., 2022). Naruto's selfie case drew attention to questions of data ownership for content created by animals. It raises issues about whether an animal can hold copyright and, if not, who should be entitled to manage or benefit from such content. It also prompts ethical questions about how any money generated through animal-created media should support the animal involved or contribute to the welfare of the species more broadly.

A second area of ethical concern relates to the researcher's positionality when studying animals on social media and their capacity to engage in critical reflexivity about how animals are represented. The ways in which researchers interpret, frame and portray animals can shape public perceptions and influence how people understand and relate to different species. Reflexive awareness is also essential to avoid projecting human assumptions, emotions, or narratives onto animals – whether through anthropomorphism or by unintentionally "speaking for" them. This concern also connects to the earlier discussion on consent: how can researchers interpret, judge or operationalise the idea of animal consent in the creation and circulation of social media content?

A third ethical dimension concerns how to balance responsibilities towards individual animals, single species or multispecies collectives. When advocating for an interspecies perspective on social media and promoting responsible business practices, we must reflect on whom we are ethically accountable to. Are researchers and practitioners accountable to an individual animal, to a species, or to multiple species within an ecosystem? The case of Cecil the lion illustrates these tensions: while the moral outcry on social media focused intensely on the fate of one individual, it also had unintended consequences for human communities and wildlife conservation efforts (see Makuyana, 2018). This example highlights the complexity of ethics when animals are taken into account.

Practical implications for responsible management

Adopting an interspecies perspective on social media has important managerial implications for organisations seeking to operate responsibly and ethically in an increasingly interconnected digital environment. First, it invites managers to broaden their understanding of stakeholders beyond human beings to include animals that may be affected – directly or indirectly – by social media content (see Tallberg et al., 2022). This expanded stakeholder perspective encourages organisations to assess how their online communications represent animals and ecosystems and how such representations may influence public perceptions, consumption behaviours and conservation outcomes. Second, this requires greater responsibility in content creation and curation. Managers should move beyond anthropomorphised or purely entertainment-driven representations of animals and instead promote narratives that highlight ecological contexts, species-specific needs and respectful human–animal relationships. Achieving this may require interdisciplinary collaboration with conservation experts, NGOs or scientists to ensure accuracy and ethical integrity, thereby reducing the risks of misrepresentation and unintended harm.

Third, interspecies social media strategies can serve as a tool for education and behavioural change. By leveraging storytelling, visual content and interactive features, organisations can raise awareness about lesser-known or threatened species and encourage responsible behaviours, such as ethical consumption, wildlife protection or support for conservation initiatives. From a managerial standpoint, this positions social media not only as a marketing channel but also as a platform for social and environmental impact. Fourth, integrating an interspecies perspective into social media management can enhance organisational legitimacy and stakeholder trust. Stakeholders increasingly expect businesses to demonstrate a genuine commitment to sustainability and responsibility not only towards other human beings but also towards animals and nature. Transparent communication of interspecies considerations – such as content guidelines, ethical standards or partnerships with conservation organisations – can strengthen organisational credibility and align corporate values with broader societal expectations.

Finally, adopting this approach requires developing internal capabilities and governance mechanisms. Managers may need to establish internal guidelines, training programmes and evaluation metrics that explicitly account for interspecies impacts. This need is further reinforced by recent developments in sustainability governance, particularly the growing recognition of biodiversity as a key dimension of corporate impact. For example, the first ISO standard addressing biodiversity, ISO 17298: Biodiversity in strategy and operations, has recently been launched by the International Organisation for Standardisation. As biodiversity becomes increasingly embedded in responsible business strategies, its implications are likely to extend to social media management, requiring approaches that are more inclusive of diverse species. This chapter repositions social media not merely as a marketing tool, but as a site of interspecies governance where animals must be recognised as stakeholders in responsible digital management.

Reflective activities inside/outside the classroom

A Revision questions that test knowledge ("what?")

- A1 According to the chapter, how has social media been conceptualised in relation to responsible business and what key limitations have been identified in this context?
- A2 What are the key elements of social media?
- A3 What are the three principles of an interspecies perspective on social media outlined in this chapter?
- A4 What is meant by an interspecies perspective on social media and what new insights does this perspective bring to discussions of social media and responsible business?

B Reflective questions that build understanding ("so what?")

- B1 Why is it important to consider the principles of an interspecies perspective on social media outlined in this chapter?
- B2 How can social media channels be leveraged to support both human wellbeing and the wellbeing of other species?
- B3 Why is it essential to move beyond an anthropocentric view of social media in discussions of responsible business?
- B4 Why should nonhuman species be recognised as key stakeholders in social media-related practices and decision-making?

C Take action ("do")

- C1 What aspects should I consider when creating social media content from an interspecies perspective?

- C2 How can I use social media to raise awareness of different species and promote respectful relationships between humans and animals?
- C3 How could an interspecies perspective on social media be applied within my university or workplace?
- C4 How might I design a social media campaign that benefits both humans and nonhuman species?

D Ideas to discuss with colleagues, friends and family ("share")

- D1 To what extent is it morally acceptable to use animals in the creation of social media content?
- D2 Can the mere representation of animals on social media be considered an interspecies perspective? Why or why not?
- D3 Are there examples in your field where an organisation has used social media in ways that align with the interspecies perspective discussed in this chapter?
- D4 Have you ever reflected on the presence of animals on social media? What are your thoughts on how they have been represented?

Key ideas

Social media

Social media can be defined as internet-based channels that allow users to opportunistically interact and selectively self-present, either in real time or asynchronously, with both broad and narrow audiences who derive value from user-generated content and the perception of interaction with others (Carr & Hayes, 2015).

An interspecies perspective on social media

An interspecies perspective views social media not only as a space for human communication but also as a site where relationships between humans and nonhuman species are shaped and carry real ethical and ecological consequences. It recognises animals and other species as relevant stakeholders whose representation, visibility and treatment are influenced by digital narratives and interactions.

Conservation marketing

Conservation marketing is the ethical application of marketing strategies, concepts and techniques to influence attitudes, perceptions and behaviours of individuals and ultimately societies, with the objective of advancing conservation goals (Wright et al., 2015).

Endangered species

An endangered species is defined as one that is in danger of extinction throughout all or a significant portion of its range (Saterson, 2013).

References

Adhikari, D. (2016). Exploring the differences between social and behavioral science. *Behavioral Development Bulletin*, 21(2), 128–135. doi:10.1037/bdb0000029.

Aichner, T., Grünfelder, M., Maurer, O., & Jegeni, D. (2021). Twenty-five years of social media: A review of social media applications and definitions from 1994 to 2019. *Cyberpsychology, Behavior and Social Networking*, 24(4), 215–222. doi:10.1089/cyber.2020.0134.

Aichner, T., & Jacob, F. (2015). Measuring the degree of corporate social media use. *International Journal of Market Research*, 57, 257–276. doi:10.2501/IJMR-2015-018.

Ballejo, F., Plaza, P. I., & Lambertucci, S. A. (2021). Framing of visual content shown on popular social media may affect viewers' attitudes to threatened species. *Scientific Reports*, 11, 13512. doi:10.1038/s41598-021-92815-7.

Bergman, J. N., Buxton, R. T., Lin, H.-Y., Lenda, M., Attinello, K., Hajdasz, A. C., Rivest, S. A., Nguyen, T. T., Cooke, S. J., & Bennett, J. R. (2022). Evaluating the benefits and risks of social media for wildlife conservation. *Facets: A Multidisciplinary Open Access Science Journal*, 7, 360–397. doi:10.1139/facets-2021-0112.

Bucci, N. (2025). Baby wombat-snatching US influencer apologises and says she was "concerned" for Australian animal. *The Guardian*. www.theguardian.com/environment/2025/mar/15/baby-wombat-snatching-us-influencer-apologises-and-says-she-was-concerned-for-australian-animal-ntwnfb.

Burns, G. L. (2015). Animals as tourism objects: Ethically refocusing relationships between tourists and wildlife. In K. Markwell (Ed.), *Animals and tourism: Understanding diverse relationships* (pp. 44–59). Channel View Publications. doi:10.21832/9781845415051-007.

Carr, C. T., & Hayes, R. A. (2015). Social media: Defining, developing and divining. *Atlantic Journal of Communication*, 23, 46–65. doi:10.1080/15456870.2015.972282.

Castelló, I., Etter, M., & Årup Nielsen, F. (2016). Strategies of legitimacy through social media: The networked strategy. *Journal of Management Studies*, 53, 402–432. doi:10.1111/joms.12145.

Chekunov, S., Stringham, O., Toomes, A., Prowse, T., & Cassey, P. (2024). Scale of unregulated international trade in Australian reptiles and amphibians. *Conservation Biology*, 38, e14355. doi:10.1111/cobi.14355.

Collard, R.-C. (2020). *Animal traffic: Lively capital in the global exotic pet trade*. Duke University Press. doi:10.2307/j.ctv15kxg0c.

Colleoni, E. (2013). CSR communication for organizational legitimacy in social media. *Corporate Communication: An International Journal*, 18, 228–248. doi:10.1108/13563281311319508.

Courchamp, F., Jaric, I., Albert, C., Meinard, Y., Ripple, W. J., & Chapron, G. (2018). The paradoxical extinction of the most charismatic animals. *PLoS Biology*, 16(4), e2003997. doi:10.1371/journal.pbio.2003997.

Daly, N. (2019). Why people risk their lives for the ultimate animal selfie. *National Geographic*. www.nationalgeographic.com/animals/article/psychology-of-why-people-enter-wild-animal-enclosures-at-zoos.

Donaldson, M. R., Burnett, N. J., Braun, D. C., Suski, C. D., Hinch, S. G., Cooke, S. J., & Kerr, J. T. (2016). Taxonomic bias and international biodiversity conservation research. *Facets*, 1(1), 105–113. doi:10.1139/facets-2016-0011.

Dwivedi, Y. K., Ismagilova, E., Hughes, D. L., Carlson, J., Filieri, R., Jacobson, J., Jain, V., Karjaluoto, H., Kefi, H., Krishen, A. S., Kumar, V., Rahman, M. M., Raman, R., Rauschnabel, P. A., Rowley, J., Salo, J., Tran, G. A., & Wang, Y. (2021). Setting the future of digital and social media marketing research: Perspectives and research propositions. *International Journal of Information Management*, 59, 102168. doi:10.1016/j.ijinfomgt.2020.102168.

Effing, R., van Hillegersberg, J., & Huibers, T. (2011). Social media and political participation: Are Facebook, Twitter and YouTube democratizing our political systems? In E. Tambouris, A. Macintosh, & H. de Bruijn (Eds), *Electronic participation: ePart 2011*. Lecture notes in computer science, vol. 6847 (pp. 25–35). Springer. doi:10.1007/978-3-642-23333-3_3.

Etter, M., Colleoni, E., Illia, L., Meggiorin, K., & D'Eugenio, A. (2018). Measuring organisational legitimacy in social media: Assessing citizens' judgments with sentiment analysis. *Business & Society*, 57(1), 60–97. doi:10.1177/0007650316683926.

Etter, M., & Vestergaard, A. (2015). Facebook and the public framing of a corporate crisis. *Corporate Communication: An International Journal*, 20, 163–177. doi:10.1108/CCIJ-10-2013-0082.

Hamilton, L., & Taylor, N. (2017). *Ethnography after humanism: Power, politics and method in multispecies research*. Palgrave Macmillan. doi:10.1057/978-1-137-53933-5.

Hunter, M. L., Le Menestrel, M., & De Bettignies, H. C. (2008). Beyond control: Crisis strategies and stakeholder media in the Danone Boycott of 2001. *Corporate Regulation Review*, 11, 335–350. doi:10.1057/crr.2008.30.

Huopalainen, A., Satama, S., & Tallberg, L. (2025). Aesthetic encounters, canines and care: New multispecies methodological avenues in organisational and business ethics. *Journal of Business Ethics*, 203, 241–257. doi:10.1007/s10551-025-06035-4.

Kaplan, A. M., & Haenlein, M. (2010). Users of the world, unite! The challenges and opportunities of social media. *Business Horizons*, 53, 59–68. doi:10.1016/j.bushor.2009.09.003.

Kidd, L. R., Gregg, E. A., Bekessy, S. A., Robinson, J. A., & Garrard, G. E. (2018). Tweeting for their lives: Visibility of threatened species on Twitter. *Journal of Nature Conservation*, 46, 106–109. doi:10.1016/j.jnc.2018.10.001.

Kozinets, R. V. (2015). Netnography. In P. H. Ang & R. Mansell (Eds), *The international encyclopedia of digital communication and society*. Wiley Online Library. doi:10.1002/9781118767771.wbiedcs067.

Kredens, C., & Vogt, C. A. (2023). A user-generated content analysis of tourists at wildlife tourism attractions. *Frontiers of Sustainable Tourism*, 2, 1090749. doi:10.3389/frsut.2023.1090749.

Lundberg, P., Vainio, A., MacMillan, D. C., Smith, R. J., Veríssimo, D., & Arponen, A. (2019). The effect of knowledge, species aesthetic appeal, familiarity and conservation need on willingness to donate. *Animal Conservation*, 22(5), 432–443. doi:10.1111/acv.12477.

Lundgaard, D., & Etter, M. (2023). Everyday talk on Twitter: Informal deliberation about (ir)responsible business conduct in social media arenas. *Business & Society*, 62(6), 1201–1247. doi:10.1177/00076503221139838.

Makuyana, P. G. (2018). Trophy hunting: To ban or not to ban? Legal pathways for Zimbabwe in the aftermath of Cecil the lion. *Journal of Sustainable Development Law and Policy*, 9(1), 133–167. doi:10.4314/jsdlp.v9i1.7.

Mkono, M. (2018). The age of digital activism in tourism: Evaluating the legacy and limitations of the Cecil anti-trophy hunting movement. *Journal of Sustainable Tourism*, 26(9), 1608–1624. doi:10.1080/09669582.2018.1489399.

Mkono, M., & Holder, A. (2019). The future of animals in tourism recreation: Social media as spaces of collective moral reflexivity. *Tourism Management Perspectives*, 29, 1–8. doi:10.1016/j.tmp.2018.10.002.

Mkono, M., & Hughes, K. (2024). A state-of-the-art-review of animals in tourism: Key debates and future directions. *Tourism Geographies*, 27(3–4), 653–665. doi:10.1080/14616688.2024.2342462.

Saterson, K. A. (2013). Government legislation and regulations in the United States. In S. A. Levin (Ed.), *Encyclopedia of biodiversity* (2nd ed., pp. 712–721). Academic Press. doi:10.1016/B978-0-12-384719-5.00186-6.

Schultz, F., Castelló, I., & Morsing, M. (2013). The construction of corporate social responsibility in network societies: A communication view. *Journal of Business Ethics*, 115(4), 681–692. doi:10.1007/s10551-013-1826-8.

Shaw, M. N., McLeod, E. M., Borrie, W. T., & Miller, K. K. (2022a). Human positioning in close-encounter photographs and the effect on public perceptions of zoo animals. *Animals*, 12(1), 11. doi:10.3390/ani12010011.

Shaw, M. N., Borrie, W. T., McLeod, E. M., & Miller, K. K. (2022b). Wildlife photos on social media: A quantitative content analysis of conservation organisations' Instagram images. *Animals*, 12(14), 1787. doi:10.3390/ani12141787.

Shaw, M., Dunn, M., Crowley, S., Owen, N., & Veríssimo, D. (2024). Using photo editing to understand the impact of species aesthetics on support for conservation. *People and Nature*, 6, 1–16. doi:10.1002/pan3.10602.

Siriwat, P., & Nijman, V. (2018). Illegal pet trade on social media as an emerging impediment to the conservation of Asian otters species. *Journal of Asia-Pacific Biodiversity*, 11(4), 469–475. doi:10.1016/j.japb.2018.09.004.

Smith, R. J., Salazar, G., Starinchak, J., Thomas-Walters, L. A., & Veríssimo, D. (2020). Social marketing and conservation. In W. J. Sutherland, P. N. M. Brotherton, Z. G. Davies, N. Ockendon, N. Pettorelli, & J. A. Vickery (Eds), *Conservation research, policy and practice*. Ecological reviews (pp. 309–322). Cambridge University Press. doi:10.1017/9781108638210.019.

Statista. (2025). *Number of internet and social media users worldwide as of October 2025*. www.statista.com/statistics/617136/digital-population-worldwide/#:~:text=Showing%20entries%201%20to%202,:%20100%25;%20height:%20auto%20.

Tallberg, L., García-Rosell, J.-C., & Haanpää, M. (2022). Human–animal relations in business and society: Advancing the feminist interpretation of stakeholder theory. *Journal of Business Ethics*, 180, 1–16. doi:10.1007/s10551-021-04840-1.

Troudet, J., Grandcolas, P., Blin, A., Vignes-Lebbe, R., & Legendre, F. (2017). Taxonomic bias in biodiversity data and societal preferences. *Scientific Reports*, 7(1), 913. doi:10.1038/s41598-017-09084-6.

van der Meer, E., Botman, S., & Eckhardt, S. (2019). I thought I saw a pussy cat: Portrayal of wild cats in friendly interactions with humans distorts perceptions and encourages interactions with wild cat species. *PLoS One*, 14(5), e0215211. doi:10.1371/journal.pone.0215211.

Williams, S. (2016, May 24). The most deadly selfie: Man killed while taking pictures with a 1.5-tonne walrus as the animal "playfully dragged him into the water." *Daily Mail*. www.dailymail.co.uk/news/peoplesdaily/article-3606379/The-deadly-selfie-Man-killed-taking-pictures-WALRUS-animal-playfully-drags-water.html.

Wright, A. J., Veríssimo, D., Pilfold, K., Parsons, E. C. M., Ventre, K., Cousins, J., Jefferson, R., Koldewey, H., Llewellyn, F., & McKinley, E. (2015). Competitive outreach in the 21st century: Why we need conservation marketing. *Ocean & Coastal Management*, 115, 41–48. doi:10.1016/j.ocecoaman.2015.06.029.

PART III

Learning with/from animals about innovation, creativity and diversity

7

ANIMALS AND INNOVATION

Introduction

As an interdisciplinary field of study, innovation has generated a substantial body of research. Innovation plays a key role across sectors such as the military, technology, healthcare and manufacturing, as well as in public policy and education (Eriksson et al., 2025). Innovation teams or roles are common in many organisations. Although innovation is usually portrayed as a human practice involving creative thinking and novel solutions that contribute to society's wellbeing, animals have also found their way into the field. Indeed, human beings have been studying animals as a source of innovation by mimicking their skills, exploring their anatomy and using them for testing purposes as part of the innovation process. Drawing inspiration from animals and other nonhumans to create sustainable and efficient solutions to meet human needs is well known as the notion of "biomimicry" – or the practice of learning from and emulating nature's forms, processes and ecosystems to address human design and technological challenges (Benyus, 1997). While this concept offers a new way of viewing animals, shifting our focus from exploiting them to learning from them, biomimicry primarily focuses on finding solutions to human challenges.

For instance, energy-efficient building designs have been inspired by the ventilation systems in termite mounds, which naturally regulate temperature (Andréen & Soar, 2023). Similarly, studying the adhesion system of a gecko's foot, renowned for its exceptional adhesive properties, has led to the creation of strong, reusable adhesives (Wang et al., 2021). The Wright brothers also studied the flight of birds to develop the first successful aeroplane. Since then, wing designs and flapping mechanisms have continued to draw influence from the study of flying animals. The study of animal movement,

DOI: 10.4324/9781003585541-11

known as animal locomotion, has been instrumental in driving groundbreaking innovations across fields such as technology, engineering, robotics and biomechanics (Biewener & Patek, 2018). There are also more troubling examples of innovations inspired by animals, often rooted in systems of animal exploitation. One remarkable case is Henry Ford's moving automobile assembly line, which was inspired by the disassembly lines used in Chicago's Swift and Company meatpacking house for slaughtering and processing animals (Henry Ford Museum of American Innovation, n.d.). The use of animals in testing cosmetic products, medications and vaccines is another example of dark innovation processes, driven by the need to prove safety or efficacy for human use.

These discussions about animals in relation to innovation portray them as passive objects, to be observed, studied, imitated or, in the worst cases, exploited for human purposes. In these narratives, little attention is given to the agency of animals in innovation processes or how such innovations could contribute to objectives beyond human needs. This tendency to instrumentalise animals is not entirely surprising, given the long-standing anthropocentric premises that have shaped innovation studies, which generally consider agency and communication as exclusively human capacities (see García-Rosell et al., 2025). As a result, animals are rarely acknowledged as contributors to processes that shape innovation, reinforcing a narrow, human-centric view that overlooks their broader ecological and social significance. This anthropocentric framing is not accidental but embedded in how innovation has historically been conceptualised, measured and rewarded. By centring economic growth, efficiency and human benefit, innovation studies have largely excluded animals as subjects of concern or as stakeholders in their own right. Drawing on the work of Nona Borgström, Anna Hielm-Björkman, Astrid Huopalainen and Linda Tallberg on multispecies wellbeing and animal participation in human-animal work, this chapter examines how innovation might be rethought through an interspecies lens. In doing so, it argues that an interspecies approach requires not simply adding animals into existing models, but fundamentally questioning the assumptions that define innovation itself.

Our expert collaborators and the case

This chapter is based on interviews with Anna Hielm-Björkman, Nona Borgström, Astrid Huopalainen and Linda Tallberg, researchers in the "People and Animal Wellbeing at Work and in Society" (PAWWS) project, funded by the Research Council of Finland. PAWWS advances understanding of multispecies wellbeing in organisations, emphasising animal participation in human–animal work. Building on approaches including posthuman, critical and One Health / One Welfare, the project challenges anthropocentric views that treat animals as tools or resources, advocating ethical coexistence and the recognition of animals as integral yet marginalised members of society. Through multidisciplinary collaboration across organisational research, veterinary medicine and social and health sciences, PAWWS examines the interconnection between animal health and wellbeing and that of humans, as well as what it means for animals to be part of human organising activities such as work. Anna and Nona work on preventive veterinary medicine and scent-detection dog research. Anna is a member of the Finnish Association of Pain Studies and the founder of the Pain and Rehabilitation Clinic at the University Animal Hospital, University of Helsinki. She launched a pioneering research initiative on scent-detection dogs and established the COVID-19 screening programme with scent-detection dogs at Helsinki Airport in 2020. Nona is a PhD researcher with DogRisk, a research group led by Anna at the University of Helsinki. Her work focuses on studying dogs trained to detect and distinguish pain. This research is conducted in collaboration with the Orton Pain Hospital in Helsinki, the DogRisk research group and the dog training organisation InnoMediDogs. Drawing on her background in social work and dog ethology, Nona contributes to the development of approaches that enhance the wellbeing of both humans and animals.

Astrid and Linda bring together posthumanism, feminist approaches and critical animal studies to explore human–animal relationships at work and "more-than-human" organising, pushing the boundaries of traditional organisational research to include multispecies and ethically engaged frameworks. Astrid and Linda are co-authors of a call to action for multispecies inclusivity published as a Commentary in *Business & Society* (Tallberg et al., 2024). Together with Lindsay Hamilton, Linda co-edited the book *The Oxford Handbook of Animal Organisation Studies* (Tallberg & Hamilton, 2022). Both Astrid and Linda have played key roles in creating an agenda for the inclusion of nonhuman animals in business and society and in advancing organisational theory by challenging human-centric paradigms. They have demonstrated a strong commitment to integrating a multispecies perspective into management education in master's and doctoral-level courses.

This chapter builds on the multidisciplinary research conducted within the PAWWS project, with particular attention to Anna and Nona's work on medical scent-detection dogs, including COVID- and pain-detection contexts. Drawing on conceptual perspectives developed through the collaboration – as well as previous work by Astrid and Linda in particular – we analyse medical scent-detection dog studies as a form of interspecies innovation, considering both human and animal wellbeing. We identify key principles to guide responsible interspecies innovation.

Dogs have been used for millennia for a wide range of odour-detecting tasks, as their ability to discriminate odours is superior to that of humans and of machines made by humans (Turunen et al., 2024). Medical detection dogs discriminate diseased samples from non-diseased samples and alert patients to upcoming disease attacks. COVID-sniffing dogs refer to dogs that were specially trained during the pandemic to scent-detect the virus, where large-scale, rapid screening of passengers was required within a short period of time (see Kantele et al., 2022; Meller et al., 2022). Pain detection dogs, on the other hand, are trained to identify physiological changes associated with oncoming pain episodes before they become severe, alerting their human partners to take proactive measures, such as medication or rest. As Anna and Nona note, early intervention by pain-detection dogs can reduce the intensity and duration of pain episodes, improving pain management and quality of life. These dogs can help reduce medication use by motivating their human partners to adopt non-medical pain management techniques, such as physiotherapy, meditation and breathing exercises, leading to a more holistic, sustainable pain management strategy. In both medical scent-detection cases, the dogs communicate through non-verbal cues. These include changes in proximity or specific behaviours, such as alerting by touching a sound device, which their human partners learn to interpret and trust, strengthening mutual understanding and reliance. While the focus is primarily on human benefits, medical scent-detection work also creates career opportunities for dogs with an exceptional sense of smell and eagerness to use their natural sniffing instincts. From this perspective, research and development on medical scent-detection dogs provides a compelling context for examining innovation processes in which humans and animals are active participants and their wellbeing is interconnected.

In the following sections, we define innovation, including social and sustainable innovation, and critique its anthropocentric limitations. We then outline the case for including animals in academic discussions of innovation, highlighting the practical, ethical and conceptual value of this inclusive approach. Drawing on the case of medical scent-detection dogs, we show how viewing animals as active participants can broaden our understanding of innovation. Finally, we call for research and organisational practices that recognise animals as both contributors to and beneficiaries of innovation, advancing more inclusive and ethically responsible management approaches that strengthen human–animal wellbeing.

Conceptual foundations: What is innovation?

In management studies, innovation is typically conceptualised as the process by which firms transform new or improved ideas into marketable solutions, encompassing varying degrees of novelty relative to their context of application, with its benefits most extensively examined within an economic framework (Montonen, 2025). Schumpeter, for example, identified innovation as the cornerstone of economic change, emphasising its transformative impact on markets and industries (Callegari & Nybakk, 2022; Martin, 2012). He not only established innovation as a key subject of inquiry but also framed it as a transformative force that shapes industries, creates wealth and transforms societies. It is no surprise that innovation has become a widely used term in modern society, deeply embedded in organisational strategies and political discourse, contributing to the rise of innovation centres across university campuses (Kahn, 2018). While Kanter (1983) defines innovation as the process of bringing new, problem-solving ideas into practice, Bessant and Tidd (2011) describe it as the process of translating ideas into valuable, practical new products, processes or services. Innovation as an outcome is understood as the introduction of new products and services. Likewise, it encompasses changes in methodology or processes aimed at improving efficiency, such as faster processing, greater throughput or reduced costs (Kahn, 2018). As highlighted by Phills et al. (2008), a process or outcome is considered an innovation if it meets two criteria. The first is novelty: While innovations need not be entirely original, they must be new to the user, context or application. The second criterion is improvement: the process or outcome must be more effective or efficient than existing alternatives. These two criteria are complemented by Kahn (2018), who suggests that innovation should also be understood as a mindset that fosters an environment in which individuals are empowered to think differently, challenge assumptions and experiment with new ideas.

Social and sustainable innovation

Although Schumpeterian theory and much of the innovation literature have primarily focused on the relationship between innovation, entrepreneurship and economic progress, there has been an attempt to extend the principles of innovation beyond economic goals. This emerging literature explores how innovation can also contribute to sustainability and social objectives, addressing environmental challenges and societal wellbeing (Eriksson et al., 2025). According to Phills et al. (2008, p. 39), social innovation refers to "a novel solution to a social problem that is more effective, efficient, sustainable, or just than existing solutions and for which the value created accrues primarily to society as a whole rather than private individuals." In this conception of innovation, value refers to the creation of benefits or cost reductions for society by addressing social needs and problems, going beyond the private gains of firms or the general economic benefits derived from market activity (Phills et al., 2008). While social and sustainable innovation share many similarities, sustainable innovation places greater emphasis on environmental goals as one of its main aims. For example, Tello and Yoon (2008, p. 165) define sustainable innovation as "the development of new products, processes, services and technologies that contribute to the development and wellbeing of human needs and institutions while respecting natural resources and regeneration capacities."

In these discussions, some attention is given to animal-related economic sectors, such as the pet industry, farming and food production (e.g. Provensi et al., 2024; Schäfer & König, 2018; Sodano, 2019). Within this scholarship, sustainability innovations are often presented as outcomes, processes and mindsets that reduce waste, improve energy efficiency and promote the responsible use of natural resources. Animal welfare is frequently included among these anticipated benefits, yet it does not necessarily improve in parallel with environmental or economic gains. For example, efforts to reduce greenhouse gas emissions in livestock production may promote indoor confinement systems that optimise feed efficiency but restrict animals' movement and "natural" behaviours (see Chapter 2). Selective breeding and faster-growing poultry strains can lower emissions per kilogram of meat produced, yet are associated with lameness and other health problems (see Chapter 4). Similarly, precision technologies that intensify monitoring and productivity may improve resource efficiency while increasing the density or control of animal bodies. In other sectors, renewable energy infrastructure, such as wind farms, has raised concerns about bird and bat mortality. When animals benefit from these innovations, it is often because their welfare aligns with the environmental and economic goals of business organisations. This alignment cannot be taken for granted and may involve trade-offs between environmental, economic and animal welfare objectives. Sustainability innovation also engages animals as sources of inspiration. Biomimicry, in particular, has played a pivotal role, with numerous sustainable solutions developed by drawing on animals, their behaviours and their adaptive capacities (García-Rosell et al., 2025).

While none of these definitions explicitly state that innovation must address human interests, they strongly imply that the satisfaction of human needs is often a key driver and indicator of innovation, whether viewed as a process, an outcome or a mindset. Indeed, the user-centric approaches and design thinking emphasised in innovation processes suggest that the goal of innovation is to create solutions that provide value to human beings (Drucker, 1985; Kahn, 2018). Similarly, human knowledge and ingenuity are seen as prerequisites for the success of innovation processes. But where are the animals? If animals have played an important role in innovation processes, why are they missing from the innovation discussion? Are innovations only important for humans, or can they also benefit animals? Can animals play an active role in the innovation process and in knowledge creation? How can animals be included in innovation, so that they are validated as both knowledge creators and beneficiaries?

Towards interspecies innovation

Based on interviews with the expert collaborators, their perspectives on the relationship between their work and innovation and our own interpretation of their contributions to the field, we identify three key principles for advancing an interspecies innovation approach in which both humans and animals are recognised as active co-creators and beneficiaries of innovation: Multispecies co-creation, the prioritisation of animal wellbeing and animal-informed consent.

Multispecies co-creation

Building on the concept of "multispecies inclusivity" (Tallberg et al., 2024), this chapter highlights the intrinsic value, perspectives and interests of animals in innovation processes. An interspecies perspective on innovation means that animals are not only seen as opportunities to access new ways of knowing, but also as partners, co-creators and beneficiaries of innovation. Interspecies innovation should foster solutions that enhance the wellbeing of humans, animals and other nonhuman species. Recognising animals as co-creators in innovation processes and relying on them for the success of these initiatives highlights their remarkable capacity for interaction and impact, as well as the extent to which humans depend on their knowledge and skills to address sustainability challenges (see García-Rosell et al., 2025). This challenges the dominant instrumental view of animals as exploitable resources in innovation processes, advocating instead for recognition of their agency and constitutive role in shaping innovative ideas and outcomes. Some of these aspects are visible in the COVID-sniffing dogs, which can be regarded as an interesting case of interspecies innovation. The dogs demonstrated their ability to identify multiple COVID-19 variants, making them particularly reliable as a non-invasive pre-screening method to save time and resources, or even as the sole testing method when other approaches were unavailable. Indeed, the validation research showed that the dogs were able to identify the positive patients 4–5 days before the PCR test was positive (Kantele et al., 2022).

The primary innovation in this case lay in leveraging dogs' exceptional olfactory skills and scent memory. Dogs possess a highly advanced sense of smell, with millions more scent receptors than humans, enabling them to detect subtle changes in volatile organic compounds emitted by individuals infected with the virus, even in pre-symptomatic or asymptomatic stages (Kantele et al., 2022; Turunen et al., 2024). Remarkably, once trained, these dogs can reliably adapt to identify new scents of interest within four to six weeks. The training process extended beyond the dogs to the researchers themselves, who had to learn to interact with them, trust the dogs' assessments and interpret the information gathered through their olfactory systems. As Anna explains, human participants initially tended to trust diagnostic scientific methods over the dogs, only to later discover that the dogs were always accurate in their evaluations, while the PCR test had lower sensitivity and specificity. But, as she reflects, how do we compare to a golden standard when the new test is so much more accurate than the golden standard? This work with COVID-sniffing dogs demonstrated their potential to contribute to public health protection, enabling rapid deployment in response to emerging threats, compared with conventional laboratory tests (see Meller et al., 2022). The initiative also highlighted unexplored opportunities to include scent-detection dogs in public health, revealing their significant role as reliable partners for the medical community. Beyond detecting COVID-19, scent-detection dogs have demonstrated remarkable accuracy in identifying diseases such as cancer, often surpassing existing technological capabilities (Meller et al., 2022; Muppidi et al., 2021). Such extraordinary abilities pave the way for the development of novel diagnostic tools and therapeutic approaches, challenging and expanding the boundaries of human-centric innovation paradigms. The insights gained from the work with COVID-sniffing dogs have informed subsequent innovative studies exploring the role of dogs in pain detection.

These examples underscore the transformative potential of involving animals as co-creators and stakeholders, enabling them to develop innovative practices and solutions. By recognising animals as stakeholders with agency in organisational contexts (see Tallberg et al., 2021a), we can cultivate more inclusive and respectful innovation processes that prioritise the wellbeing of both humans and animals. This paradigm shift expands the scope of innovation, fostering ethical and holistic practices valued not only for their human-centric benefits but also for their positive contributions to other species and the overall health of the planet. In other words, expanding innovation beyond human benefit is not only an organisational concern but a planetary one, as innovation processes shape ecological systems, species survival and multispecies futures. Nevertheless, to make innovation more ethical in relation to nonhuman species, we need to ensure that innovation processes not only safeguard the wellbeing of the animals involved but also actively promote it, making them direct beneficiaries of the process. To this end, we turn to the principles of prioritising animal wellbeing and exploring ways to secure animal-informed consent, which are discussed next.

Animal wellbeing as a priority

An innovation process that takes an interspecies perspective must secure the wellbeing of both humans and animals. To safeguard animal wellbeing, it is essential to involve animal behaviourists or veterinarians in developing guidelines that ensure animals' physical and psychological needs are met throughout the innovation process. In sniffing work, this is generally easy, as dogs love to sniff. It is like taking a child into a candy factory and asking them to taste candies. But there can be obstacles as well. In the case of pain-detection dogs, animal welfare is ensured through a carefully designed, collaborative selection process involving both the dogs and the patients participating in the study. As Nona observes, this requires working closely with guide dog schools and hospital personnel to ensure a suitable match, taking into account the dog's temperament, training and the patient's specific needs and circumstances. An important aspect concerns how dogs whose temperaments or behavioural profiles are better suited to tasks other than guide dog work can thrive in alternative roles, such as medical scent-detection. For instance, as Anna and Nona explain, some dogs may not be ideally matched to the specific demands of guide work, not because of any deficiency, but because their eagerness to engage in olfactory exploration exceeds what that role requires. In medical scent-detection, this pronounced sensory motivation becomes the key strength. This illustrates how the selection processes need to be attentive to aligning tasks with the capabilities, preferences and wellbeing of both humans and dogs involved in the research.

Another way to ensure the wellbeing of animal participants is through the education and training of the humans involved in the innovation process. For example, before starting an innovation process involving animals, the research team needs to conduct educational sessions for the human participants. These sessions should address topics such as specific behaviour, cognition and responsible handling. Training is vital in dispelling misconceptions and challenging ingrained anthropocentric attitudes towards animals. As Nona notes regarding scent-detection dogs, there is a need to "stop being alien to the animals" and instead recognise them as sentient beings with their own needs and personalities. By fostering a deeper understanding of animals through education (see Tallberg et al., 2021b), we contribute to a shift from viewing them as mere tools to appreciating them as partners and co-participants in an innovation process. Open communication and boundary setting also play a critical role in ensuring the wellbeing of animals involved in an innovation process. In the case of pain-detection dogs, as Nona explains, this includes maintaining clear communication and establishing boundaries among the various stakeholders, including the patient, healthcare professionals, researchers and the dogs themselves. It is about explicitly defining roles and responsibilities, addressing potential conflicts of interest and ensuring the dog's needs are not overshadowed by patient care or healthcare professionals' embedded practices. Researchers need to monitor the dogs' behaviour for signs of stress or burnout, assess the patients' ability to care for the dogs and ensure regular check-ups with veterinarians and other healthcare professionals. Finally, all our expert collaborators agree that formal agreements and contracts are needed to safeguard the wellbeing of both dogs and humans involved. These documents articulate ethical principles, clarify responsibilities for care and welfare and establish procedures for addressing potential ethical issues.

Animal-informed consent

Astrid and Linda's reflections on seeking consent from animals highlight an important issue in assessing animals' preferences and willingness to participate in an innovation process. Both expert collaborators comment that discussions of animal consent are emerging in the literature as a relatively new yet highly relevant area of exploration. For instance, Fennell (2022) recently proposed an innovative animal-informed consent framework, suggesting that animals can express consent or dissent through their emotions, preferences, behaviours and physical states. In the case of medical scent-detection dogs, a similar approach is used, employing specific practices to address the complex issue of obtaining consent from the animals involved in the studies.

In research with human participants, "informed consent" refers to a process in which individuals are given clear information about a study, understand its potential risks and benefits and voluntarily agree to participate. Since this kind of explicit, information-based consent is not possible with animals, we need to approach the issue differently. Animals cannot be provided with written information sheets or asked to give verbal agreement. Instead, we must carefully observe animal behaviour to gain insights into their experiences and interpret non-verbal signals, such as recognising stress through eye cues, panting, gestures, expressions and movement. Rather than offering definitive solutions, research on medical scent-detection dogs and other forms of canine work can provide valuable insights into the concept of informed consent, with direct implications for other nonhuman species involved in innovation processes.

In this regard, the use of multispecies methodologies can be viewed as a valuable tool for understanding animal perspectives and their consent (Huopalainen et al., 2026). Developing animal-centred research methods and committing to methodologies that prioritise animal wellbeing and agency are seen as key steps towards a better understanding and assessment of animal preferences and choices (Dashper, 2020). Methodological approaches such as multispecies ethnography, art-based methods, multisensory walking methods and the use of visual devices such as body cameras have been proposed as possible directions for this work (see Huopalainen et al., 2026). Additionally, prioritising non-invasive techniques is highlighted as an effective way to minimise potential harm or distress to animals involved in innovation. For instance, saliva tests to measure cortisol levels in dogs offer a non-invasive method for assessing stress, aligning with the principles of informed consent discussed by Fennell (2022).

At the same time, understanding animal participation requires attention not only to biological or medical indicators, but also to the social and organisational structures through which animals are included in innovation processes. The social sciences help us examine how knowledge about animals is produced, how categories such as "welfare" or "consent" are defined and how assumptions about agency, expertise and value shape practice (Clancy et al., 2022). These ways of thinking, acting and knowing directly influence how animal wellbeing is interpreted and prioritised in the first place. By drawing on different disciplinary perspectives, researchers can gain a more holistic understanding of animals as living subjects as well as of the human-made systems and institutional logics that organise their involvement in innovation, rather than viewing them merely as resources. This broader lens strengthens efforts to support meaningful forms of animal-informed consent. In practical terms, this means that innovation cannot be evaluated solely in terms of efficiency, scalability or economic value. Managers must ask whose wellbeing is enhanced, whose agency is recognised and whose bodies bear the costs. In this sense, interspecies innovation is not a niche concern but central to ethical management practice.

Directions for future research

In this chapter, we showed that, for the field of innovation to be more comprehensive, it needs to expand beyond human-centred and environmental frameworks to include animal perspectives. This is particularly relevant for innovation approaches framed as sustainable or responsible. Future studies can explore different ways of reimagining through an interspecies perspective, one that includes animals not merely as passive participants or resources, but as active agents, co-creators and beneficiaries. At the same time, an interspecies approach must also consider questions of access and scalability. Not all individuals can have a pain-detection dog in their home, whether for practical, economic or personal reasons. Insights gained from working with scent-detection dogs may therefore contribute to the development of technological applications, such as advanced olfactometers or other scent-detection devices, that build on animals' sensory capacities. In some cases, individuals may prefer a technical solution rather than working with a dog. This will help us acknowledge the interconnectedness of all species and embrace more inclusive, ethical and sustainable ways of imagining and organising innovation processes. An interspecies perspective on innovation invites us to consider how animals shape, influence and participate in organisational life. It challenges conventional, extractive and anthropocentric models by advocating relational, embodied and ecologically responsive innovation practices. This involves asking difficult questions: Is working together always beneficial for both people and animals? When might technological substitution be appropriate and how can it be developed without erasing the contributions of animal co-creators? What does it mean to co-organise with animals in ways that respect their agency, needs and perspectives? These questions demand empirical study of how organisations operationalise interspecies innovation, including how they measure animal wellbeing, allocate responsibility and justify technological substitution.

To advance this agenda, future research must embrace sensory, aesthetic and embodied ways of knowing and understanding that go beyond verbal and written language (see Huopalainen et al., 2026). Animals communicate through mimicking, breathing patterns, movement, sound, presence and affect and innovation studies must develop methods that can engage with these non-verbal modes of interaction. Arts-based approaches, validated animal behaviour indices, multisensory ethnography and embodied methodologies offer promising avenues for capturing the richness of multispecies innovation, but also need further development to overcome their anthropocentric limitations (see Dashper, 2020; Hielm-Björkman et al., 2003, 2009; Huopalainen et al., 2026). Moreover, future studies should consider the roles of gender dynamics, power relations and cultural worldviews in the inclusion of animals in innovation. Indigenous perspectives, which often view animals and plant life as kin and collaborators, can provide valuable insights into alternative ways of organising and innovating. These perspectives challenge dominant paradigms in innovation and open the way to more inclusive and respectful futures for both animals and humans.

Finally, future qualitative research on innovation must include diverse species and adopt methods that prioritise affirmative ethics, relational care and justice for animals (Huopalainen et al., 2026). This involves recognising the positionality and agency of animals in society and in innovation ecosystems and advocating for their inclusion not as objects or tools, but as beings with intrinsic value and rights. By embracing multispecies ethics and methodologies, innovation research can help create workplaces and societies that are truly sustainable and inclusive for all species.

Ethical implications

Including animals as active participants in innovation processes, even when their agency is recognised, raises important ethical challenges (Huopalainen et al., 2026). In particular, the inherent power imbalance between humans and animals must be acknowledged and addressed in any innovation process that seeks greater inclusivity of animals. Given the anthropocentric perspective prevailing in the field of innovation, such processes often risk perpetuating exploitation or reinforcing existing hierarchies between humans and other species. This dynamic becomes particularly evident in the healthcare sector, where animals are frequently treated as resources; for instance, pain-detection dogs are considered the property of the hospital.

While innovation processes such as those involving pain-detection dogs may aim to foster more attentive and respectful forms of collaboration, this does not in itself eliminate structural inequalities or fully rebalance power relations, as animals' bodies and behaviours continue to be mobilised in ways that primarily serve human needs. At the same time, in this specific context, scent-detection tasks build on activities that many dogs find intrinsically motivating and enjoyable, complicating a simple narrative of exploitation. Recognising animals as co-creators may therefore be understood as an important step towards challenging instrumental framings of their involvement, even if such recognition does not automatically translate into more ethical treatment.

A more animal-inclusive approach to innovation requires deconstructing prevailing hierarchies and creating spaces with reduced power imbalances. While this approach can be implemented within individual innovation processes, fostering a broader shift in the field of innovation necessitates influencing social and institutional structures. As Astrid and Linda highlighted, addressing power dynamics requires systemic change, including legal and policy reforms. This shift would involve developing animal-inclusive legislation that protects animal wellbeing and recognises their agency (Coulter, 2016). In addition, educational initiatives are essential for raising awareness of the capabilities and needs of animals across various work contexts. Open dialogue among researchers, practitioners and policymakers is crucial to ensure innovation processes that genuinely embrace animal inclusion and promote more equitable and ethical practices. For animal inclusion in sustainable innovation to be ethically defensible, there must be a shift from viewing animals as mere tools for human benefit to taking their wellbeing and needs seriously in decision-making processes. As long as animals are treated instrumentally, existing power hierarchies will remain unchanged. From this perspective, the case of the medical scent-detection dogs and the theoretical perspectives discussed in this chapter offer valuable insights for the development of a more ethical approach to innovation – one that prioritises the wellbeing, needs and agency of both animals and human beings (see Dashper, 2020; Huopalainen et al., 2026).

Practical implications for responsible management

Adopting an interspecies perspective on innovation has profound practical implications for responsible management. Building on the concept of multispecies inclusivity (Tallberg et al., 2024), this approach encourages managers and organisations to recognise animals not merely as passive subjects or tools, but as beings with intrinsic value, agency and the capacity to contribute meaningfully to innovation processes. It also invites critical reflection on how innovation processes themselves shape animals' roles, bodies and experiences. By including animals as co-creators, organisations can gain access to new ways of knowing and problem-solving that are often inaccessible through human-centred approaches alone. This shift not only enhances the wellbeing of animals but also can lead to more ethical and sustainable innovations that benefit both humans and nonhumans.

The cases of COVID-sniffing and pain-detection dogs exemplify this potential. These dogs demonstrated the ability to detect multiple variants of the virus and different types of chronic pain with remarkable accuracy. Their success was not only due to their advanced olfactory systems but also to the collaborative process, which required researchers to learn from and trust the dogs' assessments. This interspecies collaboration revealed the dogs' capacity for adaptation, learning and reliable detection, challenging traditional assumptions about expertise and authority in human–animal work. It also highlighted the underutilised role of dogs in public health, suggesting that their capabilities could be extended to other medical areas.

From a management perspective, this calls for a redefinition of who is acknowledged as a stakeholder within organisations and innovation processes. Recognising animals as stakeholders with agency calls for a reassessment of organisational values, decision-making processes and ethical responsibilities (Tallberg et al., 2021a). Responsible management, in this context, involves creating inclusive environments where animal contributions are respected and supported and where innovation is guided by principles that prioritise the wellbeing of all involved species. This includes designing humane training protocols, fostering interdisciplinary collaboration and developing policies that reflect an ethical commitment to humans, animals and other species. Embracing an interspecies perspective expands the scope of innovation beyond human-centric goals. It encourages managers to consider the broader ecological and social impacts of their work, fostering a culture of care, respect and shared benefit. Finally, business organisations aiming to foster responsible innovation should consider the practical value of interdisciplinary collaboration, especially when adopting an interspecies perspective. Engaging with experts from fields such as animal behaviour, veterinary science, organisation studies, philosophy, the arts and animal studies can significantly enrich innovation processes by introducing diverse ontological and epistemological perspectives. This kind of collaboration encourages a culture of mutual learning, where assumptions are challenged and new insights emerge through shared exploration. Taking animals seriously, therefore, plays a significant role in transforming innovation – more broadly – from a tool of growth into a site of ethical responsibility and organisational accountability.

Reflective activities inside/outside the classroom

A Revision questions that test knowledge ("what?")

- A1 According to the chapter, how has the traditional concept of innovation been defined and what limitations have been identified?
- A2 What have social innovation and sustainable innovation traditionally prioritised and how have animals been positioned within these approaches?
- A3 How does the case of pain-detection dogs challenge the prevailing understanding of social and sustainable innovation?
- A4 What is meant by interspecies innovation?

B Reflective questions that build understanding ("so what?")

- B1 Why does an interspecies approach to innovation require us to reflect on animal welfare, consent and power relations?
- B2 How can innovation practices be redesigned to respect animal agency and wellbeing?
- B3 What are the consequences of leaving anthropocentric assumptions unchallenged in innovation studies?
- B4 Why is it important to consider animals as both knowledge creators and beneficiaries of innovations?

C Take action ("do")

- C1 What specific steps could I take to promote interdisciplinary approaches that include an animal-centred perspective in my studies or workplace?
- C2 How can I raise awareness among peers or colleagues about the ethical dimensions of animal-inspired innovation?
- C3 What policies or guidelines could I advocate for in my institution to ensure responsible use of animals in research and innovation?
- C4 How might I design a project or study that benefits both humans and animals, rather than prioritising one over the other?

D Ideas to discuss with colleagues, friends and family ("share")

- D1 If innovation processes claim to be "animal-informed," how can we distinguish between improving welfare and optimising animals for productivity? For example, when innovations in milking machines are guided by cows' behaviour, is the goal truly their welfare, or merely greater production efficiency? Can such innovations genuinely be considered ethical and sustainable?
- D2 Under what conditions, if any, do you think animals should be involved in innovation processes? What safeguards would be necessary?
- D3 Are there examples in your field of work where animal-inspired design or thinking could lead to sustainable solutions?
- D4 Have you ever used or seen a product inspired by animals? What did you think about it?

Key ideas

Animal-informed consent

Animal-informed consent refers to practices that interpret behavioural, physiological and contextual signals to assess animals' willingness or reluctance to participate in human-organised activities, recognising the limits of traditional human-centred consent models (Fennell, 2022).

Biomimicry

Biomimicry refers to the practice of learning from and emulating nature's forms, processes and ecosystems to address human design and technological challenges (Benyus, 1997).

Interspecies innovation

Interspecies innovation can be defined as a process of creating solutions through the interaction of different species, where both humans and nonhumans serve as sources of inspiration, co-creators and agents influencing outcomes that benefit people, animals, nature and the planet (see García-Rosell et al., 2025).

One Health / One Welfare

A One Health / One Welfare approach is a set of multidisciplinary actions and systems that promote human health and wellbeing, animal welfare, human–animal relationships and environmental protection (Mota-Rojas et al., 2022).

Posthumanism

Posthumanism refers to a bundle of theories that question and seek to disrupt the onto-epistemological binaries that prioritise culture over nature, mind over body and humans over other animals (Huopalainen et al., 2026).

Multispecies inclusivity

Multispecies inclusivity involves adopting a perspective that recognises the intrinsic value, standpoint and interests of animals within organisations. It encompasses a wide spectrum, from considering group-specific interests to individual factors such as species-specific biology and behaviour, as well as an individual's personality, traits and experiences (Tallberg et al., 2024).

References

Andréen, D., & Soar, R. (2023). Termite-inspire metamaterials for flow-active building envelopes. *Frontiers in Materials*, 10, 1126974. doi:10.3389/fmats.2023.1126974.
Benyus, J. (1997). *Biomimicry: Innovation inspired by nature*. Harper Perennial.
Bessant, J., & Tidd, J. (2011). *Innovation and entrepreneurship* (2nd ed.). John Wiley.

Biewener, A., & Patek, S. N. (2018). *Animal locomotion*. Oxford University Press. doi:10.1093/oso/9780198743156.001.0001.

Callegari, B., & Nybakk, E. (2022). Schumpeterian theory and research on forestry innovation and entrepreneurship: The state of the art, issues and an agenda. *Forest Policy and Economics*, 138, 102720. doi:10.1016/j.forpol.2022.102720.

Clancy, C., Cooke, F., & Gorman, R. (2022). Learning with donkeys: A "more-than-human" approach to animal-assisted activities. *Society & Animals*, 32(4), 341–361. doi:10.1163/15685306-bja10103.

Coulter, K. (2016). *Animals, work and the promise of interspecies solidarity*. Palgrave Macmillan. doi:10.1057/9781137558800.

Dashper, K. (2020). Holidays with my horse: Human–horse relationships and multi-species tourism experiences. *Tourism Management Perspectives*, 34, 100678, doi:10.1016/j.tmp.2020.100678.

Drucker, P. (1985). The discipline of innovation. *Harvard Business Review*, 63, 67–73.

Eriksson, P., Montonen, T., Laine, P.-M., & Hannula, A. (2025). Introduction: Possibilities and limits of innovation management. In P. Eriksson, T. Montonen, P.-M. Laine, & A. Hannula (Eds) *Elgar encyclopedia of innovation management* (pp. xiii–xix). Edward Elgar. doi:10.4337/9781035306459.fm6.

Fennell, D. A. (2022). Animal-informed consent: Sled dog tours as asymmetric agential events. *Tourism Management*, 93, 104584. doi:10.1016/j.tourman.2022.104584.

García-Rosell, J.-C., Hannula, A., & Siltaoja, M. (2025). Non-human stakeholders. In P. Eriksson, T. Montonen, P.-M. Laine, & A. Hannula (Eds), *Elgar encyclopedia of innovation management* (pp. 249–253). Edward Elgar. doi:10.4337/9781035306459.00074.

Henry Ford Museum of American Innovation. (n.d.). Swift & Company's meat packing house, Chicago, Illinois, "Splitting backbones and final inspection of hogs," 1910–1915. Collections of The Henry Ford. www.thehenryford.org/collections-and-research/digital-collections/artifact/354536/#slide=gs-188181.

Hielm-Björkman, A. K., Kuusela, E., Liman, A., Markkola, A., Saarto, E., Huttunen, P., Leppäluoto, J., Tulamo, R. M., & Raekallio, M. (2003). Evaluation of methods for assessment of pain associated with chronic osteoarthritis in dogs. *Journal of the American Veterinary Medical Association*, 222(11), 1552–1558. doi:10.2460/javma.2003.222.1552.

Hielm-Björkman, A., Rita, H., & Tulamo, R.-M. (2009). Psychometric testing of the Helsinki chronic pain index by completion of a questionnaire in Finnish by owners of dogs with chronic signs of pain caused by osteoarthritis. *American Journal of Veterinary Research*, 70(6), 727–734. doi:10.2460/ajvr.70.6.727.

Huopalainen, A., Satama, S., & Tallberg, L. (2026). Aesthetic encounters, canines and care: New multispecies methodological avenues in organisational and business ethics. *Journal of Business Ethics*, 203, 241–257. doi:10.1007/s10551-025-06035-4.

Kahn, K. B. (2018). Understanding innovation. *Business Horizons*, 61, 453–460. doi:10.1016/j.bushor.2018.01.011.

Kantele, A., Paajanen, J., Turunen, S., Pakkanen, S. H., Patjas, A., Itkonen, L., Heiskanen, E., Lappalainen, M., Desquilbet, L., Vapalahti, O., & Hielm-Björkman, A. (2022). Scent dogs in detection of COVID-19: Triple-blinded randomised trial and operational real-life screening in airport setting. *BMJ Global Health*, 7, e008024. doi:10.1136/bmjgh-2021-008024.

Kanter, R. M. (1983). *The change masters*. Simon and Schuster.

Martin, B. R. (2012). The evolution of science policy and innovation studies. *Research Policy*, 41(7), 1219–1239. doi:10.1016/j.respol.2012.03.012.

Meller, S., Al Khatri, M. S. A., Alhammadi, H. K. (2022). Expert considerations and consensus for using dogs to detect human SARS-CoV-2-infections. *Frontiers in Medicine*, 9. doi:10.3389/fmed.2022.1015620.

Montonen, T. (2025). A broad perspective on innovation management. In P. Eriksson, T. Montonen, P.-M. Laine, & A. Hannula (Eds), *Elgar encyclopedia of innovation management* (pp. 2–6). Edward Elgar. doi:10.4337/9781035306459.00008.

Mota-Rojas, D., Monsalve, S., Lezama-García, K., Mora-Medina, P., Domínguez-Oliva, A., Ramírez-Necoechea, R., & Garcia, R. d. C. M. (2022). Animal abuse as an indicator of domestic violence: One Health, One Welfare approach. *Animals*, 12(8), 977. doi:10.3390/ani12080977.

Muppidi, S. S., Katragadda, R., Lega, J., Alford, T., Aidman, C. B., & Moore, C. (2021). A review of the efficacy of a low-cost cancer screening test using cancer sniffing canines. *Journal of Breath Research*, 15(2). doi:10.1088/1752-7163/abd07f.

Phills, J. A., Deiglmeier, K., & Miller, D. T. (2008). Rediscovering social innovation. *Stanford Social Innovation Review*, 6(4), 34–43. doi:10.48558/GBJY-GJ47.

Provensi, T., Marcon, M. L., Schmidt, J. L., Rodrigues, C., & Sehnem, S. (2024). Sustainability and innovation in the pet industry: An analysis from the perspective of ecological modernisation theory. *Revista de Administracao de Empresas*, 64(3), 1–24. doi:10.1590/S0034-759020240301x.

Schäfer, M., & König, B. (2018). The role of cooperation for sustainability innovations in the agriculture and food sector. *International Journal of Agricultural Extension*, 6(3), 65–78.

Sodano, V. (2019). Innovation trajectories and sustainability in the food system. *Sustainability*, 11(5), 1271. doi:10.3390/su11051271.

Tallberg, L., & Hamilton, L. (Eds). (2022). *The Oxford handbook of animal organization studies*. Oxford University Press. doi:10.1093/oxfordhb/9780192848185.001.0001.

Tallberg, L., García-Rosell, J.-C., & Haanpää, M. (2021a). Human–animal relations in business and society: Advancing the feminist interpretation of stakeholder theory. *Journal of Business Ethics*, 180, 1–16. doi:10.1007/s10551-021-04840-1.

Tallberg, L., Välikangas, L., & Hamilton, L. (2021b). Animal activism in the business school: Using fierce compassion for teaching critical and positive perspectives. *Management Learning*, 53(1), 55–75. doi:10.1177/13505076211044612.

Tallberg, L., Huopalainen, A., & García-Rosell, J.-C. (2024). Beyond anthropocentrism: A call to action for multispecies inclusivity. *Business & Society*, 64(7), 1263–1268. doi:10.1177/00076503241271254.

Tello, S. F., & Yoon, E. (2008). Examining drivers of sustainable innovation. *International Journal of Business Strategy*, 8(3), 164–169.

Turunen, S., Paavilainen, S., Vepsäläinen, J., & Hielm-Björkman, A. (2024). Scent detection threshold of trained dogs to eucalyptus hydrolat. *Animals*, 14(7), 1083. doi:10.3390/ani14071083.

Wang, W., Liu, Y., & Xie, Z. (2021). Gecko-like dry adhesive surfaces and their applications: A review, *18*, 1011–1044. doi:10.1007/s42235-021-00088-7.

8

ANIMALS AND CREATIVITY

Introduction

Western intellectual traditions have long privileged humans as the pinnacle of creativity and rational agency. From Aristotle to Descartes and later to Marx's industrial vision of labour, Western philosophy has consistently drawn a sharp line between human reason and nonhuman instinct. Marx's comparison of the architect and the bee in *Capital* (1867/1976) crystallises this understanding. While bees produce intricate hexagonal cells, Marx insisted that even the most unskilled architect is superior to the bee, because humans first design mentally what they later build physically. This enduring distinction between conceptual foresight and instinctive action has long shaped how societies value intelligence, innovation and creative agency, positioning nonhuman life as mechanical, unconscious and ultimately exploitable.

This anthropocentric framework has far-reaching consequences. It underpins both capitalist modes of production and contemporary discourses of innovation, which treat the natural world as raw material for human creativity and innovation (Ingold, 1983). Within this worldview, animals become instruments that enable human progress, while their own capacities for adaptation and problem-solving are ignored. Bees, for instance, are valued primarily for pollination services and honey production (Seeley, 2019; Shanahan, 2022) rather than as ecological agents whose adaptive capacities and collective intelligence shapes ecosystems. In effect, anthropocentrism erases nonhuman creativity and blinds us to the distributed, relational foundations of ecological resilience. Drawing on Janet Lowore's work on forest beekeeping in Africa, this chapter offers an alternative perspective, reframing creativity as an emergent property of interspecies collaboration and system resilience rather than an aspect of human innovation.

DOI: 10.4324/9781003585541-12

The value of an animal-centred approach to creativity and resilience

Resilience and creativity are increasingly recognised in accounts of human–animal relations (Clark, 2024; Harris & Holman Jones, 2022) and combined they offer a productive lens to unsettle pervasive anthropocentric assumptions. Originating in ecology (Holling, 1973) and extended into social-ecological systems thinking (Folke, 2006), resilience describes the biological, social, or hybrid capacity of systems to absorb disturbance and reorganise. Unlike stability, which implies equilibrium, resilience highlights adaptability and renewal, foregrounding uncertainty, transformation and feedback and privileging flexibility over control. In this way, resilience aligns with contemporary multispecies and posthuman scholarship (Haraway, 2016; Puig de la Bellacasa, 2015; Tsing, 2015), which values interdependence, improvisation and relational care as creative forces.

When it comes to honey bees, resilience is revealed through the notion of distributed intelligence. Tom Seeley's (2010, 2017, 2019) research on honey bees shows that colony survival depends not on hierarchy or command but on decentralised decision-making, communication and feedback loops. Worker bees assess foraging conditions, scout new nesting sites and reach consensus through collective behaviours. Seeley calls this process "honey-bee democracy," a form of organisation where resilience emerges from the capacity to self-regulate through shared information and adaptive responses. These processes can also be seen in social systems, especially those seeking to move beyond rigid hierarchies towards more participatory and networked forms of organisation (Lee, 2024). As we observe through the case of wild African forest bees in this chapter, this process can be linked to processes of creative resilience and "creative wilding" (Harris & Holman Jones, 2022).

Our expert collaborator

The work of our expert collaborator in the field of African forest beekeeping extends this ecological thinking. As Director of Programmes at the NGO Bees for Development, Janet Lowore has worked across several sub-Saharan African countries and the insights that emerge from her work outlined in this chapter provide a rich empirical case through which to examine the value of an animal-centred, multispecies approach to creativity, adaptation and resilience. Working with beekeepers in several countries across sub-Saharan Africa, Lowore's research (Lowore, 2020; Lowore & Bradbear, 2012, 2016; Lowore et al., 2018; Meaton et al., 2021) demonstrates how livelihoods, forests and wild bees form an intertwined socio-ecological system sustained by care, observation and adaptive practice.

Unlike industrial beekeeping, where hives are standardised, colonies are artificially bred and honey flows are maximised through artificial and technological interventions (Blacquière & Panziera, 2018; Seeley, 2019;

Shanahan, 2022), forest beekeeping is low-input and ecologically embedded. Beekeepers craft hives from locally sourced materials, place them within forest canopies and allow wild colonies to self-select. Bees are not managed or manipulated; they are collaborators whose behaviour signals environmental change. The system's resilience lies precisely in its distributed agency and feedback. Bees respond to shifting floral resources and climate patterns and beekeepers adjust their activities accordingly. This resonates with the resilience literature's emphasis on adaptive cycles (Holling, 1973).

In Zambia's Miombo forests, honey yields fluctuate with seasonal rainfall, forest fires and flowering cycles. During dry periods, when forest fires threaten nectar sources, beekeepers engage in preventive burning or selective clearing to protect habitats. Their actions are not based on rigid plans but on continuous observation of ecological signals which they learn from bees, trees and weather patterns. This attentiveness embodies what has been referred to as *response-ability* (Haraway, 2007, 2016; Hustak & Myers, 2012; Stengers, 1997) – that is, the ethical and practical capacity to respond to other beings in shared worlds.

This chapter situates these arguments within a broader shift from anthropocentric to interspecies ethics. Scholars such as Haraway (2016), Tsing (2015) and van Dooren et al. (2016) invite us to reconsider resilience not as a managerial tool but as a way of thinking about living with uncertainty. To "stay with the trouble" (Haraway, 2016) means to resist the fantasy of mastery and embrace the complexities of coexistence. In this view, African forest beekeeping is not merely an economic activity: It is an ongoing experiment in care, observation and adaptation. As they learn from bees, beekeepers do not only cultivate livelihoods, they also cultivate an affective ecology through which creativity and curiosity characterise relations with human and non-human practitioners (Hustak & Myers, 2012). The chapter contributes to the book's broader project of developing an interspecies approach to responsible management by showing how creativity and resilience emerge from collaboration across species, rather than from human control alone.

The anthropocentric foundations of creativity and moving beyond them

Conventional theories of creativity rest on anthropocentric foundations inherited from Enlightenment humanism, which positioned reason, agency and innovation as uniquely human capacities. Built on this legacy of human exceptionalism, creativity is imagined as the property of autonomous individuals acting upon a passive world. In psychology and management studies, creativity is defined as the production of ideas that are both new and useful (Boden, 2004; Csikszentmihalyi, 1997; Runco & Jaeger, 2012), while innovation refers to their market application (Schumpeter, 1934).

These definitions locate creativity within individual cognitive abilities and assume that novelty originates in human minds before flowing outward into material or social environments. Such assumptions naturalise a human-nature divide and perpetuate the belief that creativity is a uniquely human faculty directed towards transforming, improving or exploiting external worlds (Ingold, 2014). This model sustains a politics of mastery that equates progress with control over nature rather than participation with it, binding creativity to economic growth and instrumental reason. Drawing on posthumanism, Harris and Holman Jones (2022) call for a rewilding of the imagination beyond economic and educational capture. They argue that creativity is not a resource extracted from human minds but a relational current moving through more-than-human worlds. This resonates with Ingold's (2014) notion of "correspondence" and Haraway's (2016) idea of "sympoiesis" or making-with, both of which reject creativity as mastery and frame it as a distributed, ecological event. Situated within multispecies flows, creativity becomes a property of living systems, an ecological act of resilience rather than an act of individual genius.

As noted, the ecological notion of resilience (Folke, 2006; Holling, 1973) helps us to reframe creativity as an emergent property of living systems rather than a uniquely human trait, shifting attention from control to adaptation. An interspecies approach to resilience thus begins by rejecting the hierarchical ordering of agency. It recognises that all living systems, whether human, animal or ecological, maintain themselves through continuous processes of feedback, adjustment and collaboration. As Holling's (1973) model of adaptive cycles suggests, persistence depends on the capacity to reorganise and renew across scales and temporalities. For resilience to endure, diversity, redundancy and communication are essential. The same logic applies at the level of species interaction; ecosystems thrive when they are composed of multiple, interdependent agents whose actions modulate and stabilise one another. When one actor dominates, the climate crisis demonstrates that resilience collapses.

An interspecies approach to creativity and resilience

Honey bee colonies exemplify resilience through the principles of adaptation and distributed resilience (Seeley, 2010). However, in an increasingly fragile agricultural context, industrial beekeeping works against such processes. In countries such as the US (Shanahan, 2022) and Australia (Davies & Riach, 2018) millions of honey bees are transported annually to farms to pollinate food crops, exposing them to large-scale, nutritionally poor monocultures where agrochemicals and pathogens are widespread (Seeley, 2019; Shanahan, 2022). Shanahan (2022) argues that responses to the problems honey bees face often follow a predictable route, highlighting multiple interacting stressors (parasites, pathogens, poor nutrition and pesticides) to explain their causes and the social, economic and ecological implications for beekeeper livelihoods and the agri-food system. She argues further that this narrative does not consider where the underlying stressors come from – that is, the expansion of industrial agriculture to feed an expanding human population.

Seeley's (2010) research on honey bees offers a useful counterpoint to such narratives through which colony survival depends on decentralised decision-making and collective sense-making. Worker bees gather and interpret environmental information, communicate through waggle dances and build consensus through iterative feedback loops. When swarming to find a new home, scout bees explore potential sites and communicate their findings to the colony. Through recruitment dances, dissent and co-sensing, they collectively select the best nest location without any central oversight. This process, which Seeley (2010) describes as "honeybee democracy," is a paradigmatic case of self-organisation and adaptive governance.

Crucially, resilience here does not emerge from stability but from flexibility. Colonies adjust foraging strategies in response to floral diversity and seasonal shifts; they regulate hive temperature and humidity through coordinated activity, reallocating labour dynamically according to need. When environmental conditions deteriorate, colonies reorganise their workforce and resource distribution rather than simply persisting unchanged. This capacity for dynamic adaptation is what makes honey bee societies remarkably resilient. It also mirrors the principles of social-ecological resilience articulated by Folke (2006), whereby systems endure by remaining open to transformation.

Seeley's (2017, 2019) notion of "Darwinian Beekeeping" provides a valuable theoretical link for analysing African forest beekeeping practices. It articulates a management philosophy grounded in evolutionary biology, advocating for beekeeping methods that enable honey bees to express their natural behavioural and physiological adaptations. Rather than prioritising intensive manipulation and maximised yield in the manner of industrial beekeeping, Darwinian Beekeeping emphasises the maintenance of colony health and resilience through alignment with evolved life-histories and species-specific traits. This entails practices such as maintaining locally adapted bee populations, using smaller and more natural hive structures, spacing colonies to mimic wild nesting and permitting natural reproductive swarming processes. By minimising human-centric stressors and respecting the ecological and evolutionary dispositions of bees, beekeepers can foster self-regulating colonies better equipped to withstand environmental pressures.

In challenging human-centred management theories that assume resilience depends on top-down coordination or leadership, Seeley's (2010, 2017, 2019) work demonstrates that resilience arises from the ability of multiple agents to sense, communicate and act together. When translated to human social and ecological contexts, this suggests that resilience is fostered not by command-and-control-like structures, but by networks of reciprocity, feedback and co-adaptation.

Forest beekeeping and resilience through co-adaptation

Janet Lowore's studies of African forest beekeeping (Lowore, 2020; Lowore & Bradbear, 2012, 2016; Lowore et al., 2018; Meaton et al., 2021) offer a striking illustration of these principles in practice. Across sub-Saharan Africa, beekeeping is conducted in close relation to wild bee populations that inhabit natural forest environments. Hives are constructed from locally sourced materials and suspended in trees where bee colonies self-select. Beekeepers do not artificially breed queens or control colony genetics as they do in industrial systems (Blacquière & Panziera, 2018; Seeley, 2019; Shanahan, 2022), they rely on bees' adaptive capacities to establish and maintain colonies.

This low-input, extensive system generates resilience through distributed agency and ecological embeddedness. Because it depends on wild bees rather than managed colonies, it preserves genetic diversity and ecological complexity. Beekeepers adapt their practices to the bees' rhythms, monitoring flowering cycles, rainfall patterns and fire risks, adjusting harvesting times and forest activities accordingly. As Lowore notes in our interview:

> I definitely think beekeepers have a very heightened awareness of nature. I mean, they just become good observers of the seasons, the weather, flowering patterns and other species. So as a beekeeper, you're constantly thinking about what's flowering and you might look at a plant that you haven't seen before and think, oh, is that good for bees?

Observation and responsiveness, as these insights suggest, are the foundations of resilience in such systems. Much like beekeepers:

> Bees collect data from their environment about what is flowering, in what abundance, where it is and is it a good resource. And they have to communicate that information to the rest of their colony… if there's a lime tree that's in flower in June or July, in a very, very short time, bees will have found that lime tree and will have recruited a workforce to collect the nectar.

This reciprocal relationship between bees, beekeepers and forests creates feedback loops that align ecological and economic stability. When bees thrive, forest ecosystems are pollinated and regenerated; when forests remain healthy, nectar flows are maintained, supporting both bees and human livelihoods. This synergy transforms what might otherwise be a subsistence practice into an integrated socio-ecological system. Forest beekeeping thus exemplifies what we could refer to as *sympoietic* resilience, that which is generated through *"making-with"* others (Haraway, 2016). The survival of human communities and nonhuman species thus depends not on control, but on cultivating mutual awareness and responsiveness.

Ingold's (2014) concept of "correspondence" – the unfolding of life through reciprocal response – offers a useful lens for interpreting these dynamics. For Ingold, living beings do not impose form upon a passive world; they participate in processes of becoming through ongoing engagement. Creativity, in this view, is identical with responsiveness. In African forests, bees do not design their nests from abstract blueprints, they build them through tactile improvisation, responding and adapting to material actions and social cues from forest beekeepers and dialogues with landscapes. The adaptive cycles observed in African forest beekeeping thus mirror Holling's (1973) ecological model of growth, conservation, release and renewal. Bee colonies expand and contract in relation to floral abundance and climatic stress, while forests undergo parallel cycles of regeneration and disturbance. Beekeepers' livelihoods are woven into these rhythms through honey harvesting during peak flowering periods, tree protection during the dry season and hive maintenance through variable conditions. Rather than seeking equilibrium, these systems embrace variability as the source of renewal.

Fire management provides a clear example of this adaptive logic. In Zambia's Miombo forests, mass tree flowering occurs at the hottest and driest time of year, when the risk of bushfires is greatest. Before this season arrives, beekeepers engage proactively in preventative burning and selective clearing to safeguard both nectar sources and hives. As Janet explained, such actions draw on intimate ecological knowledge, reflecting a deep understanding of the interdependence between bees, vegetation and fire regimes. These practices exemplify how human and nonhuman actors co-produce endurance through processes of co-adaptation. Honey bees adjust foraging strategies in response to climatic variability, mobilising their workforce during mass flowering events and democratically selecting new nesting sites through collective decision-making (Seeley, 2010). Beekeepers reciprocate by modifying hive placement, adjusting harvesting schedules and by coordinating fire-prevention efforts in line with seasonal and ecological cues. In this way, wild bees and beekeepers co-create resilient livelihoods, their relationship grounded in mutual responsiveness and ecological awareness. For managers, this case challenges the idea that resilience can be engineered through control. Instead, it suggests that long-term stability depends on learning to observe, adapt and collaborate within living systems.

Creative resilience and creative wilding

Such practices embody a form of creative resilience through interdependence. This contrasts with industrial apiculture, where standardisation and control diminish diversity and flexibility, eroding creativity (Seeley, 2019; Shanahan, 2022). Harris and Holman Jones (2022) describe such processes as "creative wilding," moments when vitality escapes enclosure and re-enters the messy, adaptive life of the world. African forest beekeeping exemplifies this wild creativity, demonstrating how humans and nonhumans ensure survival through shared, mutual care. The forest itself becomes an actor in creative ecology, shaping how hives are placed and how flowering cycles are read.

Resilience also re-centres the temporal and ethical dimensions of creativity. As Puig de la Bellacasa (2015) reminds us, creativity unfolds through the slow, attentive labour and care that sustains life rather than disrupting it. Creativity and resilience are intertwined expressions of life's adaptive capacity, distributed across species and environments rather than confined to human intention. Integrating resilience theory with posthuman accounts of creativity moves us from innovation as extraction to co-creation as endurance, a reorientation from producing the new to sustaining the conditions that allow life itself to continue. These *"reciprocal captures"* (Stengers, 2010) define the fabric of ecological life, with each participant shaping and being shaped by the capacities of others.

This co-adaptive dynamic challenges the industrial notion that resilience can be engineered. Analysis of forest beekeeping in Zambia and Ethiopia by Lowore and colleagues (Meaton et al., 2021) demonstrates that externally driven attempts to "modernise" beekeeping are unsustainable. Such interventions rest on an implicit technological determinism that equates modern equipment with improvement, but which frequently undermine ecologically embedded systems that are both economically rational and environmentally resilient. Forest hives constructed from bark, bamboo or hollow logs are low-cost, locally made and well adapted to forest conditions. Modern hives, by contrast, require capital investment, regular inspection and intensive management programmes in order for their high cost to be recovered. Technological substitution, as a result, often erodes the social and ecological relations that underpin resilient livelihoods. Resilience cannot be transferred or replicated, it must be cultivated through situated relations of care (Puig de la Bellacasa, 2015).

Gender, knowledge and distributed care

The resilience of forest beekeeping systems also depends on their social configurations. Women's participation is especially significant in regions where wild beekeeping is compatible with domestic and agricultural responsibilities (Lowore, 2021; Meaton et al., 2021). Because wild bees require minimal daily management, beekeeping offers flexible opportunities for income generation, ecological stewardship and knowledge transmission. This inclusivity enhances resilience by diversifying livelihoods and embedding ecological care across social networks, enhancing ecological and economic interdependence.

From an interspecies perspective, this social diversification parallels biological diversity in supporting adaptive capacity. Just as genetic variety strengthens colonies against disease, the inclusion of multiple forms of human participation strengthens social resilience. Knowledge exchange between generations, genders and communities thus functions as a feedback mechanism transmitting vital information through relational networks. Care, in this context, becomes a mode of communication, linking human and nonhuman actors in the collective project of survival.

Learning with bees

To adopt an interspecies approach to resilience and creativity is to confront the epistemological challenge of how humans can learn *from*, rather than merely *about*, nonhuman others. The concept of *response-ability* (Haraway, 2007, 2016; Stengers, 1997) offers useful guidance here. Learning with bees requires humility, patience and openness to nonhumans and recognition of the role of more-than-human perception and decision-making. Hustak and Myers (2012, p. 106) explain:

> What is at stake in this involutionary approach is a theory of ecological relationality that takes seriously organisms' practices, their inventions and experiments crafting interspecies lives and worlds. This is an ecology inspired by a feminist ethic of "response-ability" in which questions of species difference are always conjugated with… an affective ecology in which creativity and curiosity characterize the experimental forms of life.

This reorientation has broader implications for ecological irresponsibility and sustainability science. In mainstream sustainability discourses aligned with the SDGs, research often isolates human and nonhuman variables, measuring resilience in quantitative terms such as yield, biodiversity or income. While these metrics are valuable, they risk obscuring the lived, relational and reciprocal qualities that constitute action on the ground.

Insights from African forest beekeeping resonate with other contexts where humans and bees co-adapt. In the Italian Alps, Zola (2024, p. 170) describes migratory beekeeping in which hives are moved by beekeepers (who she terms "*multispecies specialists*") between mountain and lowland environments in response to climatic variability, integrating human improvisation with bees' environmental sensitivity. Similarly, Maderson and Elsner-Adams (2023) document caring and resilient practices among small-scale beekeepers in Europe who prioritise bee welfare and ecological regeneration over productivity.

Placing African forest systems in dialogue with these other cases shows that interspecies resilience is not culturally or geographically confined. What differs are the forms of knowledge and care through which life is expressed. In the Global North, "*natural beekeeping*" movements represent a partial rediscovery of this logic, often as lifestyle choice rather than economic necessity. In African contexts, by contrast, multispecies care is a condition of survival. Resilience is not optional but foundational.

Final reflections

While industrial systems treat resilience as a technical outcome rooted in human exceptionalism (Harris & Holman Jones, 2022), interspecies systems emphasise that ethical practice is sustained relationally. This distinction is crucial because it shapes how nonhuman life is valued. When bees are framed merely as service providers, their wellbeing becomes subordinate to human objectives. When recognised as collaborators, ethical responsibility extends across species lines. Haraway's (2016) call to "*make kin*" provides a framework for this revaluation, acknowledging interdependence and acting with care towards those who enable our survival. In forest beekeeping, making kin involves safeguarding forests, respecting bees' autonomy and aligning human livelihoods with ecological cycles. It is a practice of *response-ability* in which creativity, resilience and ethics are inseparable.

An interspecies perspective reframes sustainability by showing that resilience is not a resource to be engineered but a relationship grounded in co-creation, observation and feedback. Bees and beekeepers illustrate this vividly; their collaboration not only produces honey and income but also enhances ecological stability and biodiversity. Seeley's (2010, 2019) work reminds us that the wisdom of the hive lies in decentralisation, relinquishing mastery and learning to live with complexity. In a time of accelerating climate change and ecological collapse, such humility is indispensable. Building resilient futures requires learning from species like bees, whose survival depends on cooperation rather than control. African forest beekeeping demonstrates that resilience emerges not through technological innovation or managerial authority, but through shared, adaptive life in more-than-human worlds. In Seeley's terms, this is to give bees back their "*evolutionary freedom*." In Haraway's terms it is a call to "*stay with the trouble*." For both, creativity is central to understanding how resilience operates across and within multispecies environments.

Directions for future research

African forest beekeeping demonstrates that adaptive capacity is distributed across humans, bees and environments, sustained through observation and interdependence rather than intervention or control. Yet while the interspecies dimensions of resilience are increasingly recognised in theory, empirical research remains limited. Future scholarship must expand on the foundations outlined in this chapter to examine how creative adaptation operates across diverse ecological, cultural and economic contexts.

A first priority is comparative research. African forest beekeeping offers one vivid example but similar principles operate in other multispecies systems. Comparative studies could explore how ecological pressures and governance regimes shape interspecies resilience. Contrasting low-input forest systems in the Global South with intensive commercial beekeeping in the Global North would reveal how differing management logics of control versus collaboration affect biodiversity, genetic diversity and long-term system health. Similar analysis could be extended to agroforestry or pastoral agriculture, tracing how resilience is produced through relationships rather than technologies. These practices perform what Harris and Holman Jones (2022) call *research-creation*, the making of knowledge through situated, more-than-human participation, illuminating a creative ethics that refuses extractive innovation in favour of diversity, adaptability and care.

There is also a need to develop methodologies that incorporate nonhuman perspectives. Conventional resilience research relies heavily on human-centred data, overlooking the perceptual and behavioural contributions of nonhumans. Integrating ecological observation, sensor technologies and ethological insight with ethnography could capture multispecies interactions more fully (Kirksey & Helmreich, 2010; van Dooren et al., 2016). Tracking bee foraging patterns alongside beekeeper narratives or combining remote sensing of forest health with participatory field diaries would enrich understanding and challenge hierarchies that privilege human ways of knowing.

Because resilience unfolds through cycles of disturbance and renewal (Folke, 2006; Holling, 1973), longitudinal research is needed to trace how interspecies relationships evolve across seasons and generations. In African contexts, this could involve following communities and colonies through climatic variability to document how rainfall patterns, flowering cycles and fire regimes influence adaptation. Such studies would show how resilience is maintained, lost or transformed amid shifting environmental baselines. Future work should also address gendered and intergenerational dimensions, examining how gendered knowledge and labour contribute to adaptive capacity thus deepening understanding of this socio-ecological process.

Following Haraway (2016) and Puig de la Bellacasa (2015), resilience should also be treated as an ethical relation, raising questions about whose care and stability is prioritised and how humans might act responsibly within systems where nonhuman agency is central. Transdisciplinary collaboration across ecology, anthropology, philosophy and policy can thus move beyond instrumental notions to imagine governance as co-management with nature and a more inclusive, creative and enduring basis for adaptation.

Ethical implications

Recognising animals as co-creators of resilience carries profound ethical consequences. In conventional apiculture, bees are managed as instruments of productivity; they are bred, transported and medicated to serve human ends (Shanahan, 2022). This instrumental logic mirrors broader patterns of industrial agriculture (Akhtar, 2012; Wallace, 2016), where nonhuman lives are valued primarily for their utility. By contrast, African forest beekeeping and other multispecies systems foreground "*co-dependence*" and "*care*" as the foundations of adaptive capacity. Here, bees are collaborators rather than service providers and resilience arises from reciprocal attentiveness rather than control.

This shift entails what Haraway (2007, 2016) and Stengers (1997) call *response-ability*, the ethical and epistemological capacity to notice and respond to other beings without appropriating their agency. To act responsibly within multispecies systems is not to "speak for" nonhumans but to remain open to their signals and needs. Beekeepers who interpret swarming behaviour, floral patterns or colony health as environmental data exemplify this attentiveness. Their practice embodies an ethics of listening and care that resists the instrumentalisation of nature. However, recognising nonhuman agency also exposes tensions. Focusing on honey bees can inadvertently marginalise other pollinators and perpetuate honey bee exceptionalism. Moreover, as Janet herself notes in our interview, bee societies are also efficient and sometimes ruthless, which poses questions about whether, how and to what extent their organisational forms should inspire human governance. Ethical engagement with interspecies resilience therefore demands humility and an acknowledgment that coexistence involves conflict, dependency and negotiation rather than harmony alone.

Ultimately, interspecies ethics reframes responsibility as co-flourishing. It calls for practices that maintain the capacity of all human and nonhuman participants to adapt and thrive. In this sense, resilience is not only ecological but moral, outlining a commitment to sustaining the relationships and conditions that make shared futures possible.

Practical implications for responsible management

Combining resilience thinking (Folke, 2006; Holling, 1973) and multispecies ethics (Haraway, 2016; Tsing, 2015) offers a powerful challenge to conventional business and management paradigms. Mainstream management theory has long been dominated by human-centred assumptions about control, prediction and optimisation. Organisations are often viewed as machines designed to maximise efficiency and profit, with nonhuman life and ecological systems reduced to externalities or "resources" to be managed. Yet, as the evidence from African forest beekeeping shows, creativity and long-term business sustainability depend not on domination but on responsiveness, diversity and relational care.

Reframing management through a multispecies lens begins by recognising that all economic activity is embedded within living systems. Forest beekeeping exemplifies this principle, sustaining livelihoods precisely because it aligns human practices with ecological processes. Beekeepers' success depends on the health of forests, the adaptability of wild bees and the ecosystems that support both. In business terms, this model foregrounds *co-production* and *co-creation* rather than extraction. Bees' labour cannot be owned or controlled; it must be respected and collaborated with. Similarly, sustainable enterprises must acknowledge that resilience arises through networks of interdependence, including those that extend beyond the human.

From this perspective, responsible management is not a technical function but an ethical practice of relationship building and maintenance. Following Stengers (1997) and Haraway (2016), responsible management requires the cultivation of *response-ability*, responding to other species as stakeholders in shared worlds. This does not require fully adopting the perspective of bees, but recognising the worlds they help to shape, the connections they sustain and the forms of collaboration they make possible (Viveiros de Castro, 2004; Zola, 2024). Managers, like beekeepers, must learn to observe, listen and adapt to signals from the wider environment, whether they be ecological, social or technological. This requires moving from dominant command-and-control type hierarchies towards decentralised, adaptive forms of governance. Seeley's (2010) model of "honeybee democracy" is instructive in this sense, suggesting that the most resilient systems are those in which information flows freely, decisions are distributed and feedback is rapidly integrated.

Just as bee colonies maintain overlapping roles and genetic variation to buffer against shocks, companies can foster resilience through diverse teams, flexible supply chains and multiple knowledge systems. While efficiency-driven optimisation often strips that which enables survival under stress, care as an operational principle is also required (Maderson & Elsner-Adams, 2023; Puig de la Bellacasa, 2015). In multispecies systems, care is not sentimental, it is structural, emphasising the maintenance work required to keep relationships viable. For businesses, this translates into embedding long-term ecological and social considerations into core decision-making. Companies that cultivate care-based relationships with employees, suppliers, ecosystems and communities develop stronger adaptive capacity than those pursuing short-term returns. This insight aligns with emerging discourses in sustainability and corporate governance, where stakeholder and regenerative business models are replacing extractive paradigms (Konietzko et al., 2023).

Management can be understood similarly as a practice of stewardship (Martin & Ishihara, 2024). In African forest beekeeping, human activity supports ecological processes without dictating them. Similarly, responsible leaders must facilitate the conditions for creativity and adaptation rather than imposing rigid structures. This approach parallels Ingold's (2014) conception of "correspondence," where form emerges through ongoing engagement rather than design. Businesses that embrace stewardship cultivate an internal culture of learning, experimentation and feedback – qualities essential to navigating uncertainty. A multispecies approach also expands the boundaries of who counts as a "stakeholder" (Tallberg et al., 2024). Conventional models only recognise human employees, investors and customers, while treating nature as backdrop. Yet the resilience of supply chains, raw materials and even markets depends on nonhuman actors: soil organisms, pollinators, forests, water systems.

Governance frameworks that fail to account for these relationships perpetuate ecological vulnerability. Integrating nonhuman stakeholders into management systems through biodiversity accounting, ecosystem partnership models or regenerative certification redefines responsibility in profoundly ethical terms. In the age of ecological crisis, responsible management requires abandoning the illusion of mastery and embracing interdependence as a source of innovation and endurance. The lesson from bees and beekeepers is clear: Sustainability is achieved not by maximising control, but by cultivating the conditions for others to thrive. The future of business resilience will depend on managers who can think, decide and act creatively *with* and not simply *on behalf of* the more-than-human world.

Reflective activities inside/outside the classroom

A Revision questions that test knowledge ("what?")

- A1 How does Marx's comparison of the architect and the bee shape dominant understandings of creativity?
- A2 What distinguishes forest beekeeping from industrial beekeeping systems?
- A3 How do bees contribute to resilience through decentralised decision-making?
- A4 Why are externally driven modernisation interventions in African beekeeping often unsustainable?

B Reflective questions that build understanding ("so what?")

- B1 What does responsible management mean when resilience is understood as a multispecies process?
- B2 How can care and attentiveness be operationalised in business practice?
- B3 Can resilience involve conflict or loss and, if so, how might this shape leadership and innovation?
- B4 How might interspecies perspectives change the way we design business practices, technologies, organisations or policies?

C Take action ("do")

- C1 Spend 20 minutes observing a nonhuman system – a beehive, garden, tree canopy or patch of soil. Note how it responds to environmental change, disturbance or interaction. What forms of adaptation, cooperation or creativity do you observe? Reflect on how these dynamics might inform organisational learning or teamwork.
- C2 Choose a familiar product or service, such as honey, coffee or a smartphone, and map all the human and nonhuman contributors involved in its production. Identify the flows of energy, labour and care. Ask: whose work is visible and whose is hidden? What happens to resilience when certain relationships are neglected or exploited?
- C3 Select a routine within an organisation or classroom, such as a meeting format, supply-chain process, or learning assessment. Reimagine it by including nonhuman stakeholders (ecosystems, materials, species). How does this shift priorities, ethics and measures of success?
- C4 Compare an industrial beekeeping system with a forest beekeeping model. Identify where control dominates and where collaboration dominates. What are the implications for resilience?

D Ideas to discuss with colleagues, friends and family ("share")

- D1 Do animals have agency? What examples suggest that nonhumans make decisions, adapt or innovate?
- D2 How should ethical responsibility be distributed across species in shared ecosystems?
- D3 What might "resilience" mean if defined not only in human economic terms but in ecological or multispecies terms?
- D4 Can humans learn *with* nonhumans rather than simply *about* them?

Key ideas

Creative wilding

Creative wilding describes moments when life escapes managerial containment and re-enters the open-ended improvisation of ecological systems. It highlights how vitality emerges through unruly, multispecies interactions that defy efficiency or control. In forest beekeeping, wilding appears in bees' autonomous adaptations and in human practices that nurture, rather than suppress, this generative unpredictability (Harris & Holman Jones, 2022).

Honeybee democracy

Honeybee democracy describes how colonies make decisions through decentralised sensing, feedback and collective evaluation rather than hierarchy or command. Scout bees propose options, recruit supporters and reach consensus through iterative dances. This distributed intelligence reveals resilience as a collaborative, multispecies process where adaptive governance emerges from shared signals rather than singular authority (Seeley, 2010).

Interspecies resilience

This describes the capacity of socio-ecological systems to endure and adapt through collaborative interactions between human and nonhuman actors. Rather than relying on control or optimisation, resilience emerges from diversity, distributed intelligence and mutual responsiveness across species (Folke, 2006; Haraway, 2016).

Sympoiesis

Sympoiesis (or "making-with") rejects the idea of self-contained systems, emphasising instead that life emerges through collective processes of co-creation. It highlights how humans, animals and environments continuously shape one another, offering a framework for understanding innovation, survival and spirituality as collaborative and interdependent (Haraway, 2016).

Correspondence

Correspondence describes the unfolding of life through reciprocal response rather than imposed design. For Ingold, creativity emerges through ongoing engagement with materials, environments and other beings. It reframes innovation as participation in processes of becoming rather than the execution of pre-formed plans (Ingold, 2014).

References

Akhtar, A. (2012). *Animals and public health: Why treating animals better is critical to human welfare*. Springer. doi:10.1057/9780230358522.

Blacquière, T., & Panziera, D. (2018). A plea for use of honeybees' natural resilience in beekeeping, *Bee World*, 95(2), 34–38. doi:10.1080/0005772X.2018.1430999.

Boden, M. A. (2004). *The creative mind: Myths and mechanisms*. Routledge. doi:10.4324/9780203508527.

Clark, J. N. (2024). Human–animal connections: Expanding and cross-worlding relational approaches to resilience. *Environmental Sociology*, 10(2), 147–161. doi:10.1080/23251042.2023.2295098.

Csikszentmihalyi, M. (1997). *Creativity: Flow and the psychology of discovery and invention*. Harper Perennial.

Davies, O., & Riach, K. (2018). From manstream measuring to multispecies sustainability? A gendered reading of bee-ing sustainable. *Gender, Work & Organization*, 26 (3), 246–266. doi:10.1111/gwao.12245.

Folke, C. (2006). Resilience: The emergence of a perspective for social-ecological systems analyses. *Global Environmental Change*, 16(3), 253–267. doi:10.1016/j.gloenvcha.2006.04.002.

Haraway, D. J. (2007). *When species meet*. University of Minnesota Press.

Haraway, D. J. (2016). *Staying with the trouble: Making kin in the Chthulucene*. Duke University Press. doi:10.2307/j.ctv11cw25q.

Harris, D., & Holman Jones, S. (2022). A manifesto for posthuman creativity studies. *Qualitative Inquiry*, 28(5), 522–530. doi:10.1177/10778004211066632.

Holling, C. S. (1973). Resilience and stability of ecological systems. *Annual Review of Ecology and Systematics*, 4, 1–23. doi:10.1146/annurev.es.04.110173.000245.

Hustak, C., & Myers, N. (2012). Involutionary momentum: Affective ecologies and the science of plant/insect encounters. *Differences*, 23(3), 74–118. doi:10.1215/10407391-1892907.

Ingold, T. (1983). The architect and the bee: Reflections on the work of animals and men. *Man*, 18(1), 1–20. doi:10.2307/2801762.

Ingold, T. (2014). The creativity of undergoing. *Pragmatics & Cognition*, 22(1), 124–139. doi:10.1075/pc.22.1.07ing.

Kirksey, S. E., & Helmreich, S. (2010). The emergence of multispecies ethnography. *Cultural Anthropology*, 25(4), 545–576. doi:10.1111/j.1548-1360.2010.01069.x.

Konietzko, J., Das, A., & Bocken, N. (2023). Towards regenerative business models: A necessary shift? *Sustainable Production and Consumption*, 38, 372–388. doi:10.1016/j.spc.2023.04.014.

Lee, M. Y. (2024). Enacting decentralised authority: The practices and limits of moving beyond hierarchy. *Administrative Science Quarterly*, 69(3), 791–833. doi:10.1177/00018392241257372.

Lowore, J. (2020). Understanding the livelihood implications of reliable honey trade in the Miombo woodlands in Zambia. *Frontiers in Forests and Global Change*, 3. doi:10.3389/ffgc.2020.00028.

Lowore, J. (2021). *Forest beekeeping in Zambia: Analysing the nexus of sustainable forest management and commercial honey trade*. PhD thesis, University of Huddersfield.

Lowore, J., & Bradbear, N. (2012). Extensive beekeeping. *Bees for Development Journal*, 103, 3–5.

Lowore, J., & Bradbear, N. (2016). Honey trade conserves forests of Ethiopia. *Bees for Development Journal*, 121, 10–13.

Lowore, J., Meaton, J., & Wood, A. (2018). African forest honey: An overlooked NTFP with potential to support livelihoods and forests. *Environmental Management*, 62 (1), 15–28. doi:10.1007/s00267-018-1015-8.

Maderson, S., & Elsner-Adams, E. (2023). Beekeeping, stewardship and multispecies care in rural contexts. *Sociologia Ruralis*, 64(2), 202–221. doi:10.1111/soru.12457.

Martin, C., & Ishihara, H. (2024). Establishing a typology for stewardship: a nexus of opportunity for organisational and environmental management. *Sustainability Nexus Forum*, 32, 1. doi:10.1007/s00550-024-00543-z.

Marx, K. (1867/1976). *Capital: A critique of political economy* (Vol. I). Penguin Classics.

Meaton, J., Lowore, J., & Wood, A. (2021). Assessing value-chain interventions in Zambian and Ethiopian forest beekeeping systems. *Business Strategy & Development*, 4(2), 159–169. doi:10.1002/bsd2.136.

Puig de la Bellacasa, M. (2015). Making time for soil: Technoscientific futurity and the pace of care. *Social Studies of Science*, 45(5), 691–716. doi:10.1177/0306312715599851.
Runco, M. A., & Jaeger, G. J. (2012). The standard definition of creativity. *Creativity Research Journal*, 24(1), 92–96. doi:10.1080/10400419.2012.650092.
Schumpeter, J. A. (1934). *The theory of economic development*. Harvard University Press.
Seeley, T. D. (2010). *Honeybee democracy*. Princeton University Press.
Seeley, T. D. (2017). Darwinian beekeeping: An evolutionary approach to apiculture. *American Bee Journal*, 277–282.
Seeley, T. D. (2019). *The lives of bees: The untold story of the honeybee in the wild*. Princeton University Press. doi:10.1515/9780691189383.
Shanahan, M. (2022). Honeybees and industrial agriculture: What researchers are missing. *Journal of Insect Science*, 22(1), 4. doi:10.1093/jisesa/ieab090.
Stengers, I. (1997). *Power and invention: Situating science*. University of Minnesota Press.
Stengers, I. (2010). *Cosmopolitics I*. University of Minnesota Press.
Tallberg, L., Huopalainen, A., & García-Rosell, J. C. (2024). Beyond anthropocentrism: A call to action for multispecies inclusivity. *Business & Society*, 00076503241271254.
Tsing, A. L. (2015). *The mushroom at the end of the world: On the possibility of life in capitalist ruins*. Princeton University Press. doi:10.1515/9781400873548.
van Dooren, T., Kirksey, S., & Münster, U. (2016). Multispecies studies: Cultivating arts of attentiveness. *Environmental Humanities*, 8(1), 1–23. doi:10.1215/22011919-3527695.
Viveiros de Castro, E. (2004). Exchanging perspectives: The transformation of objects into subjects in Amerindian ontologies. *Common Knowledge*, 10(3), 463–484. doi:10.1215/0961754X-10-3-463.
Wallace, R. (2016). *Big farms make big flu*. Monthly Review Press.
Zola, L. (2024). More than human altitudes: Exploring assemblages of bees, flowers and people in the Italian Alps. *The Oriental Anthropologist*, 24(1), 166–178. doi:10.1177/0972558X241230656.

9

ANIMALS AND DIVERSITY, EQUITY AND INCLUSION

Introduction

Diversity, equity and inclusion – DEI (or EDI in the UK) – is now a central concern across business, education and government (Everett & Hill, 2025). Though often bundled together, each term has a distinct meaning. Diversity refers to the presence of differences like identity, experience and perspective within a given setting; equity involves building fair systems that recognise and accommodate structural inequalities; while inclusion fosters environments where all individuals can participate fully (Arsel et al., 2022). Together, DEI initiatives aim to recognise difference, remove barriers and promote belonging. DEI has become a normative institutional logic, guiding expectations and policies about how organisations handle demographic diversity and ethical responsibilities (Hellerstedt et al., 2024). For instance, businesses may focus on recruiting diverse leadership or supporting LGBTQ+ staff. Yet, drawing on Carlos Monterrubio's work on donkeys in rural Mexico, this chapter argues that DEI remains deeply anthropocentric and proposes an interspecies framework that extends justice across species lines.

Mainstream DEI frameworks take the "human" as the sole subject of justice, rendering other beings invisible (Jackson, 2020; Kim, 2015). In theory and practice, DEI assumes that only people can be agents and beneficiaries of justice. This approach excludes animals, even though we are all embedded in systems where human–animal relationships shape everyday life (Tallberg et al., 2024).

This exclusion reflects and reinforces "speciesism." This is a form of discrimination based on species membership that – like racism or sexism – justifies domination and limits whose interests are acknowledged (Cudworth, 2014). Speciesism thus deserves serious academic attention alongside race, gender and class as part of wider power analyses. Ignoring speciesism as a kind of critical category in its own right weakens all our DEI efforts by

DOI: 10.4324/9781003585541-13

ignoring key relationships and dependencies and – in turn – limiting how justice is understood, taught and implemented.

An animal-centred DEI framework broadens ethical horizons and reveals the multispecies interdependencies shaping organisations. It encourages reflection on how systems of power – race, gender, class AND species – intersect in shaping marginalisation. Scholars in feminist political economy and critical animal studies argue that animals are not passive tools, but active participants whose labour and vulnerability are systematically overlooked (Narayanan, 2025). Whether transporting goods, offering companionship or supporting livelihoods, animals contribute to human systems but remain excluded from recognition, rights and protections (Taylor & Sutton, 2018).

Recognising species as a category of analysis goes beyond adding another axis to DEI. Rather, it challenges our assumptions about inclusion and justice. If animals are embedded in the same systems that marginalise human groups, then DEI must extend beyond human difference (Waverley, 2024). This demands a fundamental rethinking of justice itself, as intersectional, relational and more-than-human.

Donkeys are a paradigmatic case for interspecies DEI for several reasons. Historically, they have played a central role in rural economies, particularly in Global Majority contexts, where they support agriculture, transport and informal livelihoods (Rodrigues et al., 2021). Their positioning is closely tied to structural inequalities, as they are most commonly used by marginalised communities and stigmatised as symbols of poverty or backwardness and left out of policy, academic debates and development frameworks (Clancy et al., 2022a). Importantly, their marginalisation has global consequences: In Mexico, for example, donkeys are not only exploited but are progressively disappearing as economic and policy priorities shift, revealing how exclusion operates through neglect as well as abuse (Monterrubio, 2025).

Donkeys thus offer a powerful lens for interspecies DEI. As Watson et al. (2020) and Monterrubio (2025) note, they are economically vital but culturally stigmatised: That is, they are more often viewed as emblems of poverty than contributors to sustainability. Compared with horses or oxen, they are lower in both cultural and economic status. Watson et al. (2020) highlight how donkeys are structurally marginalised due to their association with disadvantaged human communities, forming a species-specific pattern of invisibility and neglect.

This chapter contributes to the book's broader project of developing an interspecies approach to responsible management by showing how DEI frameworks can and must expand to include animals as participants in organisational justice. Mexico offers a rich context for exploring these dynamics. In rural regions, donkeys support agriculture, tourism and transport, yet are excluded from policy and heritage discourses. Monterrubio (2025) shows how donkeys are framed as rustic or backward despite their economic importance. This mirrors class, race and rural exclusion. Donkey-related work is often tied to poor, Indigenous and working-class groups – already marginalised by standard DEI frameworks. In this way, the marginalisation of donkeys cannot be separated from that of the people who depend on them. As Hribal (2012) argues, animals like donkeys are co-workers and economic actors, whose exclusion exposes structural injustice. As such, while this chapter focuses empirically on Mexico, the dynamics observed are not unique to this context. Research across parts of Africa and Asia reveal similar patterns of invisibility, stigmatisation and intersecting inequalities, where animal labour is central to livelihoods yet excluded from policy and recognition (e.g. Geiger, 2024; Watson et al., 2020). Read together, these cases point to transnational patterns in how species, class and territory intersect, rather than a series of isolated local problems. For organisations operating in tourism, agriculture and rural development, for example, these patterns are not abstract. They shape labour systems, supply chains and inclusion policies and therefore require managerial attention.

Introduction to our expert collaborator

This chapter takes as its starting point the work of Dr. Carlos Monterrubio, based at the Autonomous University of the State of Mexico. Trained in tourism studies in Mexico and the UK, he has an interdisciplinary profile spanning tourism, leisure, heritage, sport and cultural sociology. A consistent focus in his work is marginalisation, or how certain groups, places and practices are devalued or excluded within dominant systems. This critical lens led him to examine the experiences of animals – especially equids (horses, donkeys and mules) – in tourism and rural life. His work on horses explores their idealisation and the gendered, cultural dynamics surrounding them (e.g. Monterrubio et al., 2023, 2024a), but he has recently turned attention to donkeys, whose marginalisation is even deeper.

Carlos' research has helped establish animals as serious subjects of inquiry within tourism and heritage studies. His writing introduces concepts like stigma, care and representation, shedding light on how equids are framed within social systems. Of particular interest here is his work investigating how animal labour, emotion and visibility are shaped by power structures including class, gender and colonial histories. For example, his study of the Mexican national sport Charreria shows how participating horses are commodified as symbols of cultural identity while suffering poor treatment and low welfare, reflecting tensions around class and rural labour (e.g. Monterrubio et al., 2024b).

Methodologically, Carlos' work is grounded in qualitative, multispecies ethnography. Through long-term fieldwork, interviews and close attention to everyday interactions between humans and other animals, his research captures how labour, care and marginalisation are lived and negotiated across species. This approach is particularly valuable for exploring interspecies DEI relations, as it makes visible forms of agency, vulnerability and inequality that remain inaccessible through survey-based or purely human-centred methods.

Our interview with Carlos surfaced several themes, including the symbolic and material marginalisation of donkeys, the interplay between species and social difference and possibilities for more ethical interspecies relationships. He spoke about how donkeys are both relied upon and degraded and how tourism can pose both risks and opportunities for reimagining human–animal relations.

Carlos' work expands DEI's conceptual boundaries and demonstrates how interspecies perspectives illuminate questions of power, visibility and care. His critical insight and empirical depth enrich this chapter, helping us push towards more inclusive, relational and multispecies justice in tourism and organisational life.

Conceptual foundations of diversity, equity and inclusion

To meaningfully consider how animals might be integrated into DEI theory and practice, it is important to trace its conceptual roots and evolution. Such reflection reveals three key developments: A shift from legal compliance to structural justice and belonging, the growing role of intersectionality and the emerging backlash that challenges DEI's legitimacy. These foundations illuminate DEI's potential, limitations – particularly its anthropocentrism – and open the door to imagining more-than-human inclusion.

Historical origins and academic evolution

DEI originated in the civil rights movements of the 1960s and reforms like the US Civil Rights Act (1964), the UK's Race Relations Act (1965) and Sex Discrimination Act (1975), which aimed to eliminate systemic inequities in employment, education and public life (Park et al., 2025). Initially centred on legal compliance, DEI evolved into a broader organisational and ethical framework. Contemporary definitions emphasise diversity as representation, equity as fair access and inclusion as active participation and belonging (Adamson et al., 2021; Triana et al., 2021).

By the 1980s, scholarship began moving beyond legalistic models, exploring cultural and strategic paradigms. Researchers shifted focus from demographic representation to outcomes such as innovation, leadership and engagement (Van Bommel et al., 2024). Critics soon noted that DEI often prioritised surface-level diversity and economic benefit, while overlooking deeper structural exclusions and power imbalances (Adamson et al., 2021; Köllen, 2021).

From representation to belonging and intersectionality

A key pivot in DEI thought emerged from the distinction between diversity and inclusion. Mor Barak (2022) defines inclusion as feeling valued and empowered to participate fully in decision-making (Triana et al., 2021).

Others (e.g. Ferdman, 2017; Roberson, 2006) linked inclusion to engagement and performance, adding psychological depth to DEI. More critical scholars argue that inclusion frameworks often conceal exclusion, allowing organisations to claim inclusivity while ignoring migrant and minority experiences (Ahmed & Rogers, 2017). These critiques urge closer scrutiny of who is genuinely heard.

As a result, intersectionality, first popularised by Crenshaw (1991), has become central to DEI, showing how overlapping identities and systems – race, gender, class, ability – compound marginalisation. Organisational scholars apply this lens to challenge frameworks that treat identity categories as discrete (Kossek et al., 2025). In practice, this means recognising that inequalities cannot be fully understood if race, gender or class are considered separately. Justice-oriented scholars therefore argue that DEI must engage with institutional structures, not only individual bias (Fitzsimmons et al., 2023).

However, intersectionality's Anglo-American framing limits its global reach. For example, Akotirene (2019) critiques its co-optation by Eurocentric perspectives and reclaims its grounding in Afro-Latin feminist struggles. Similarly, Ndlovu-Gatsheni (2015) and Lugones (2010) emphasise coloniality as inseparable from the entanglement of race, class and gender, showing how DEI discourses often neglect legacies of colonial domination. These perspectives stress that intersectionality must reflect local histories of resistance, not only US or European legal debates. Without this, DEI risks reproducing Global North frameworks ill-fitted to postcolonial realities.

Backlash and reframing

In North America and parts of Europe, DEI now faces intensifying backlash manifesting in legislative rollbacks, bans on DEI-related training and funding in universities and vocal opposition from political figures who frame it as divisive or ideologically driven (Iyer, 2022). This resistance – often rooted in perceived threats to dominant groups – has narrowed space for critical work. In response, some organisations adopt safer terms like "belonging" or "inclusive excellence" (Park et al., 2025). This uncertain climate underscores the urgency of clarifying the aims of DEI and examining which actors and concerns remain outside its scope. While intended to soften opposition, scholars warn that such depoliticised language risks diluting DEI's transformative potential and reducing it to branding. In other words, these more palatable formulations strip away the structural critique at the heart of equity work, shifting focus from systemic change to performative inclusion (Adamson et al., 2021; Aguinis et al., 2025; Fitzsimmons et al., 2023). Compounding this dilution is the tendency – across scholarship and practice – to privilege diversity over equity and inclusion. As Hellerstedt et al. (2024) note, this emphasis is driven by the ease of measuring representation and its alignment with organisational imperatives to demonstrate profitability, market competitiveness and innovation through visible diversity metrics. In practice, this translates into diversity initiatives that prioritise demographic headcounts – such as gender or ethnicity on boards of companies and NGOs – over deeper efforts to dismantle systemic barriers, inadvertently masking exclusion and worsening structural inequalities.

While DEI faces constriction through backlash and depoliticisation in the Global North, in other parts of the world, the challenge is not resistance, but gaining an institutional foothold. In many Global Majority contexts, DEI initiatives often remain nascent or marginal, struggling for legitimacy within local governance frameworks, academic institutions and civil society. The issue isn't backlash but recognition. However, Mkono (2019) argues that reimagining DEI in these regions demands more than transposing Northern templates – it must take account of colonial legacies and structural inequities by grounding initiatives in situated struggles and epistemologies.

Across these debates – around belonging, intersectionality and backlash – DEI remains overwhelmingly anthropocentric. Dominant frameworks focus exclusively on human identity and experience, sidelining the interdependencies between people and other species. This conceptual blind spot sets the stage for rethinking DEI through a more-than-human, interspecies lens.

Towards an interspecies approach to DEI

Diversity, equity and inclusion (DEI) agendas shape contemporary discourse across corporate governance, education, policy and development. Their language informs hiring, funding and institutional reform. Yet despite this reach, DEI remains focused almost exclusively on humans – albeit people who may be marginalised and misrecognised. In theory and practice, DEI rarely acknowledges how human lives, institutions and systems are deeply entangled with those of animals (Mkono & Hughes, 2025; Riggs et al., 2021).

This silence reflects persistent human exceptionalism, where animals are viewed as property, labouring bodies or symbolic resources rather than stakeholders in justice. Legal and policy regimes reinforce this exclusion by classifying animals as objects, erasing their social presence (Blattner et al., 2020; Cudworth, 2014). As a result, shared human–animal precarity is overlooked, even where labour and livelihood are inseparably linked. In much of Europe, debates about animal welfare tend to focus on companion animals or wildlife conservation, leaving working animals largely outside ethical and policy concern (Wadham & Dashper, 2024). The Mexican case disrupts this framing by foregrounding donkeys as fundamental socioeconomic actors, whose labour sustains households, tourism and rural infrastructure and whose exclusion therefore raises questions of justice rather than charity.

Donkeys illustrate this contradiction vividly. In many Global Majority contexts, they remain essential for transporting goods, fetching water, supporting tourism and generating income for marginalised households. In rural

Mexico, donkeys form part of formal and informal economies, yet are still framed as backward relics (Monterrubio, 2025). In India's brick kilns, they labour under dangerous, unrecognised conditions (Watson et al., 2020). Globally, an estimated 50 million donkeys contribute to human livelihoods – though this figure includes both working equids and those kept for companionship or other non-labour purposes (Norris et al., 2021).

Yet donkeys (and other animals) are absent from inclusion audits, equity policies or justice frameworks. Their exclusion is not incidental but symptomatic of DEI's anthropocentrism. Universalist language often "smuggles in" the assumption that justice begins and ends with humans (Donaldson & Kymlicka, 2011). To ignore animals whose labour sustains communities which DEI claims to serve is to miss the entangled realities of inequality.

In what follows, by centring animals, we expose a structural oversight of DEI. Our specific focus on donkeys highlights how speciesism intersects with racism, sexism, class and colonial legacies and why a less human-centred approach is essential if DEI is to become genuinely relational, just and transformative.

Structural speciesism and the limits of inclusion

The exclusion of animals from DEI reflects what Cudworth (2014) and Blattner et al. (2020) term structural speciesism: the assumption, built into legal and cultural frameworks, that animals are property rather than participants. Donkeys cannot have contracts, sit on worker committees or be litigants in court. Yet they labour, suffer, relate and resist. Their contributions often blur lines between production and care, complicating dominant categories of inclusion rooted in human agency, consent and metrics.

Monterrubio (2025) highlights how donkeys in Mexican tourism are framed as quaint symbols of heritage – pulling carts in festivals, appearing in photo ops – but this framing obscures exploitative dynamics. These animals are expected to work tirelessly under poor conditions, without shade or rest, for the entertainment of typically urban or foreign tourists. Their human companions, often from rural and Indigenous backgrounds, face parallel marginalisation. The exclusion of animals and marginalised people here as elsewhere is not parallel but interwoven – entangled through mutual dependency and structural disregard (Jackson, 2020; Ko & Ko, 2017).

This dynamic is echoed in East Africa and parts of Asia, where donkeys transport goods to market, sustaining subsistence economies under significant strain (Geiger et al., 2021; Watson et al., 2022). Welfare interventions that target "bad owners" risk reproducing neoliberal tropes of individual blame, ignoring broader political economies of land loss, water scarcity and gendered labour divisions. As Perlo (2022) and Kopnina and Gjerris (2015) caution, welfare framed as individual responsibility risks reinforcing neoliberal narratives of blame rather than addressing systemic injustice.

Epistemic justice and posthuman reframing

DEI's human-centric design makes it ill-equipped to accommodate animal agency. Where animals' capacities don't fit existing rubrics – performance metrics, representation or consent – they are excluded not due to irrelevance but because dominant models lack the tools to perceive them. A more inclusive DEI must engage what we call "epistemic justice." This refers to the need to recognise alternative ways of knowing and engaging with the world, including forms of communication and contribution that fall outside standard human norms.

Posthuman thinkers (e.g. Braidotti, 2019; Jammaers & Huopalainen, 2023) argue for a shift from individualistic frameworks towards relational ethics, valuing interconnectedness across species. This demands attunement to non-verbal, embodied forms of agency – such as animal resistance and responsiveness. Watson and Clancy's (2025) work in Romanian donkey therapy centres shows how donkeys intuitively respond to children's emotional states, approaching withdrawn children and remaining still with anxious ones. Similarly, Clancy et al. (2022b) note that donkeys communicate through muscle tension, moving their ears or withdrawal – gestures that convey boundaries, preferences and affective ties. These acts express intentionality, but defy easy categorisation within DEI evaluation models. Such interactions challenge the assumption that voice, consent and recognition must occur through human language. Inclusion, in this reframed sense, becomes a matter of acknowledging the agency of beings whose modes of participation exceed verbal communication and organisational measurement.

Narayanan (2025) extends this perspective, arguing that animals are knowledge producers within food and labour systems. Their embodied expertise – anticipating terrain, adjusting pace, refusing routes – is shaped through experience and attunement, not mere instinct. This insight reframes inclusion from charity to co-production, pressing DEI to move beyond managing suffering towards enabling structural justice. The question is no longer how animals can be accommodated within DEI, but rather how DEI itself must transform to include those whose labour, knowledge and care are routinely sidelined.

Intersectionality: Class, gender, caste – and species

As we outlined in the earlier section on "conceptual foundations," intersectionality has long been central to DEI, exposing how overlapping systems – race, gender, class, caste – compound marginalisation. But this lens often stops at the species boundary. Recognising species as a further axis of inequality offers deeper insight into labour, care and exclusion in shared human–animal spaces.

Carlos' work reveals how donkey care in rural Mexico is feminised and devalued: Women and girls perform unpaid, emotionally skilled work, while donkeys are feminised through symbolic associations with docility and domesticity (Monterrubio, 2025). These gendered portrayals render both the labour and agency of carers and animals invisible. Rather than separate domains, human and animal feminisation intertwine within broader regimes of reproductive labour.

Geiger et al.'s (2023) work in Ethiopia affirms this co-marginalisation, showing how donkeys become "women's animals." Their labour is vital yet undervalued, mirroring the precarity of women's work. Watson et al. (2020) trace similar patterns in India, where donkey-owning families – often Dalit (lower caste), Muslim or migrant – face shared stigma, exclusion and instability. These relationships form complex entanglements, often affective and embodied, which enmesh humans and other creatures within specific social settings (Sayers et al., 2019). The work undertaken by marginalised people and animals is frequently rooted in everyday practices rather than formal institutions, rendering it largely invisible to wage structures and policy frameworks (Taylor & Sutton, 2018).

Such insights foreground the need to rethink labour as not solely a human concern but as deeply interwoven with the lives of other species. Donkeys are not proxies for human suffering. Rather, they are social actors in their own right, constrained but also capable of resistance, expression and contribution (Kim, 2015; Ko & Ko, 2017). For DEI to fulfil its promise of delivering a fairer world for all, it must expand to account for how species intersects with race, gender, class and coloniality, especially in sectors like tourism and development.

Ethical reparation and relational futures

Within organisations, DEI is often shaped by what can be audited – metrics, representation, performance. But important questions remain: Who gets left out by these standards? And what does inclusion mean when someone's identity or way of being doesn't fit into official categories?

Throughout the chapter, we've seen how donkeys (like other working animals) are excluded from DEI frameworks not due to irrelevance, but because existing systems can't accommodate relational, affective and nonhuman forms of labour. Carlos' work (e.g. Monterrubio, 2025) challenges this gap by proposing a reimagining of animals' roles in tourism and development, positioning them as vulnerable contributors in shared multispecies contexts. Instead of focusing only on welfare, he encourages us to recognise animal labour through accountability, responsiveness and respect across species lines. This involves questioning cultural narratives that depict donkeys as dirty, stubborn or expendable and imagining more inclusive policies and practices – ones where animals are treated as stakeholders in collective futures.

Such a shift requires a deeper epistemic transformation, one grounded in a more-than-human ethics of care (Riggs et al., 2021; Waverley, 2024). Animals are not just objects of concern but relational agents whose signals, rhythms and needs warrant attention. This calls for attunement, reciprocity and responsiveness, not as lofty ideals, but as practical ways to honour animal agency. It means adjusting workflows, noticing non-verbal cues and valuing forms of labour beyond efficiency.

If DEI is serious about justice, it must interrogate exclusions affecting both human and nonhuman contributors. Donkeys (and other animals) complicate simplified ideas of agency and consent. By focusing on relational interdependence, DEI can evolve into a multispecies practice: one that values diverse ways of working and communicating and expands moral consideration to those traditionally ignored. In doing so, justice becomes less about numbers and more about how care, labour and recognition are ethically sustained across species boundaries.

Directions for future research

Future research must take seriously the ethical and analytical consequences of excluding animals from DEI frameworks. Rather than simply adding species to existing diversity agendas, scholars should interrogate how human exceptionalism underpins current understandings of justice, labour and inclusion. An interspecies approach requires both conceptual reframing and empirical inquiry – centring the lives, contributions and marginalisation of animals like donkeys. Research should move beyond anthropocentric models and explore four interconnected areas, namely intersectionality, labour, methodology and structural justice.

One key area of inquiry is the intersection of species with other axes of oppression. Interspecies intersectionality recognises that animals are often marginalised alongside already-excluded human communities – such as Indigenous or working-class populations (Monterrubio, 2025; Watson et al., 2020). Donkeys, for example, are typically cared for by women and children in rural Mexico and India, whose own labour is undervalued and unrecognised. Their exclusion is not parallel but relational (Sayers et al., 2019). Future research could trace how species, gender, class and race interact to produce shared forms of precarity across different industries and contexts. This relational approach challenges siloed understandings of injustice and offers new tools for examining how power operates across human–animal boundaries (Kim, 2015; Mkono & Hughes, 2025).

A second area concerns multispecies methodologies and epistemic inclusion. If animals are to be recognised as stakeholders in DEI, researchers must find ways to listen across species boundaries. Donkeys communicate through movement, proximity, refusal and affective attunement (Clancy et al., 2022b; Watson & Clancy, 2025). Methodologies such as multispecies ethnography, embodied observation and participatory research with animal carers offer promising paths forward (Braidotti, 2019; Tomlinson, 2024). Future studies could explore how to treat animals as knowers and respondents – not merely as objects of care or concern. This shift raises deeper questions about what counts as voice, presence and recognition in justice-oriented research (Clancy et al., 2022b; Jammaers & Huopalainen, 2023).

Finally, future research might explore how DEI can move from welfare to structural reparation. Welfare approaches tend to frame animals as victims in need of care. A reparative approach asks harder questions: What systems render certain lives disposable? What policies or practices would recognise donkeys not only as suffering beings, but as historical participants in human economies? The above idea of ethical reparation invites scholars to examine how justice might be reimagined across tourism, development and education in ways that include – not obscure – animal labour, vulnerability and contribution (Blattner et al., 2020; Cudworth, 2014; Perlo, 2022).

Together, these areas chart a research agenda that challenges the boundaries of DEI and extends justice beyond the human. Donkeys, as both workers and symbols, help illuminate the multispecies exclusions embedded in our institutions. Future studies should follow their lead.

Ethical implications

As this chapter has shown, the absence of animals from DEI conversations isn't just a simple oversight – it reflects a deeper ethical oversight rooted in the belief that only humans matter (Donaldson & Kymlicka, 2011; Kim, 2015). Donkeys, in particular, help reveal this problem. They play vital roles in rural economies, tourism and everyday survival, yet are treated as symbols of poverty or dismissed entirely (Monterrubio, 2025; Watson et al., 2020). This disconnect forces us to ask important questions: Who is seen as having a voice? Whose work is valued? And whose needs are recognised?

Donkeys push us to rethink the idea of justice – not as something that belongs to humans alone, but as something we share across species. Carlos Monterrubio's research shows that donkeys and their human handlers – often women, racially minoritised people or families facing economic hardship – are excluded together. Animals are treated as tools, while their caregivers are treated as invisible (Cudworth, 2014). These injustices are deeply connected. To fully understand inclusion, DEI must respond to both.

The argument doesn't stop with donkeys but applies to other animals too. If we believe inclusion means actively taking part, then animals who work, have emotions, resist control and form relationships should also be taken seriously from an ethical standpoint (Hribal, 2012; Taylor & Sutton, 2018). At the moment, most research tends to focus narrowly on animal welfare – mainly on reducing harm within systems built by humans. But we need a broader approach that looks at how these systems themselves treat animals as objects or property. This means challenging deep-rooted beliefs about who owns what, what is considered valuable and who deserves to have a voice in institutions. In summary, animals shouldn't just be seen as helpless beings we take care of. They are part of our society and our politics and their involvement forces us to rethink what inclusion really means and who gets included in our decision-making spaces (Blattner et al., 2020).

Practical implications for responsible management

For managers in tourism, heritage or farming contexts, inclusion must be reconceived not as an exclusively human undertaking but as a multispecies endeavour. This reframing will not be without controversy. DEI initiatives already face backlash in many sectors, often accused of overreach, ideological bias or diverting focus from core business goals (Iyer, 2022; Park et al., 2025). Introducing animals into the inclusion frame may heighten these tensions, provoking scepticism about whether organisations are being asked to do too much, or pushing ethical boundaries too far. In a sociopolitical climate where the term "wokeism" is frequently weaponised – used to dismiss progressive reform as excessive or ideologically driven – the idea of multispecies inclusion risks being labelled as another symptom of cultural overcorrection. Yet this reaction, while rhetorically potent, often obscures the structural questions at hand. Rather than diluting DEI's intent, a multispecies lens may expose the deeper architectures of marginalisation that entangle humans and animals alike. It invites managers not to do "more" in a superficial sense, but to think differently about who counts, why and how organisational justice might operate across interdependent lives.

One immediate implication is the need to expand stakeholder definitions. If animals like donkeys are active participants in labour systems, then they must be recognised as stakeholders with interests, histories and vulnerabilities. As Tallberg et al. (2024) argue, multispecies inclusivity demands abandoning anthropocentric models and acknowledging animals as more than background entities. In practice, this means building mechanisms to evaluate their wellbeing structurally, not just individually. Organisations might, for example, include working animals in welfare audits, consult with communities and carers about the roles animals play in local economies or develop planning processes that account for shared wellbeing across species lines.

A second implication concerns co-marginalisation. The stigma directed at working animals is often mirrored in the social positioning of their human caregivers – typically women, older people, migrants or marginalised ethnic groups (Kim, 2015; Mkono & Hughes, 2025). Responsible management cannot address exclusion piecemeal. Rather, it must consider how symbolic hierarchies mark both animals and humans as dirty, backward or improper. Co-workers and managers need to incorporate intersectional analysis into their strategies – recognising how species, gender, class and ethnicity interact to shape experiences of organisational legitimacy and access. This might mean revising outreach and training practices, building interspecies capacity through community engagement, or disrupting narratives that pathologise non-mainstream relations between humans and animals.

Third, rather than treating animals as abstract ethical dilemmas, responsible organisations must interrogate how their operations rely on animal labour – often hidden, affective and essential. From festivals and brick kilns to therapeutic care farms, donkeys contribute economically and socially, yet their roles remain unspoken in policy frameworks (Watson et al., 2020). Management education should therefore integrate perspectives from critical animal studies, helping future leaders analyse power, acknowledge structural speciesism and design inclusive organisational policy that doesn't relegate animals to the margins.

Finally, these shifts call for epistemic humility. As Jammaers and Huopalainen (2023) suggest, meaningful inclusion demands attentiveness to forms of communication that cross species boundaries. Donkeys express distress, cooperation, refusal and pleasure through embodied cues and responsible managers must learn to interpret these signals with care. Cultivating attunement to animal behaviour not only transforms how work is organised, it also challenges dominant managerial norms that prioritise control and productivity. Training staff to recognise and respond to animal communication, embedding responsiveness in organisational culture and resisting assumptions of mastery are practical steps towards a more just and relational future.

In short, interspecies DEI reframes responsible management as an ethical and situated practice; one that insists on acknowledging the lives that often sustain our communities and organisations, even when the individuals in question speak in ways we're not used to hearing. This chapter therefore repositions DEI not as a purely human rights project, but as a multispecies framework for justice that responsible organisations can no longer ignore.

Reflective activities inside/outside the classroom

A Revision questions that test knowledge ("what?")

- A1 What do the letters D-E-I stand for?
- A2 How is DEI usually applied in businesses and other organisations?
- A3 Why are animals often left out of DEI conversations?
- A4 Why are donkeys presented as a useful case for thinking about interspecies DEI?

B Reflective questions that build understanding ("so what?")

- B1 Why might it be a problem to only think about humans when we talk about fairness or inclusion?

- B2 How are donkeys and the people who work with them sometimes excluded in similar ways?
- B3 What would it mean to treat donkeys not just as animals, but as workers with needs and relationships?
- B4 How might donkeys "communicate" their feelings, even without words?

C Take action ("do")

- C1 Choose an industry like farming, tourism or delivery. Where do animals help humans in this industry? Create a simple drawing or chart that shows what they do and how they might be affected by that work.
- C2 Pick an animal you're interested in (e.g. donkey, horse, elephant). Make a poster, TikTok or short slide presentation to explain how that animal helps people – and what kind of support or protection they might need.
- C3 Imagine you are designing a fair workplace that includes animals. Write three rules or values your organisation would follow to treat animals more justly.
- C4 Watch a short video of donkeys working (in tourism, farming or therapy). Write 100–150 words about what you notice. How are they treated? Do they seem tired, cared for, ignored? What questions do you have about their lives and work?

D Ideas to discuss with colleagues, friends and family ("share")

- D1 Share a story or post that shows a working animal (e.g. donkey, therapy dog, police horse). How is the animal framed – helper, hero, tool, symbol? What's missing from the narrative?
- D2 Ask a friend or family member whether they think animals can be included in ideas like fairness or justice? Why or why not? What would need to change for that to happen?
- D3 Talk to someone about how animals are seen in different cultures. For example, are donkeys seen as smart, stubborn, sacred, helpful? How do those views affect how animals are treated?
- D4 Think about your future job or career. Are animals part of it in any way? If yes, how could they be included more fairly? If not, what does this chapter make you think about your responsibilities to animals in general?

Key ideas

Speciesism

A form of discrimination based on species, where human lives are valued more than animal lives. Like racism or sexism, it justifies inequality by treating some beings as less worthy than others (Cudworth, 2014; Riggs et al., 2021).

Interspecies intersectionality

Oppressions like racism, classism, colonialism, ableism and speciesism compound each other and affect both humans and animals. Justice requires addressing these shared and overlapping inequalities rather than treating human and animal struggles as separate concerns (Monterrubio, 2025; Watson et al., 2020).

Marginalisation

The process by which individuals and/or groups are ignored or devalued by those in power. People often look down on donkeys, as they are seen as "backward" or "cheap." This means they and their human carers tend to be sidelined in policy and practice (Geiger, 2024).

Structural exclusion

Animals are left out of DEI not by accident but by design. Legal systems treat them as property and inclusion frameworks assume only humans matter. This structural exclusion hides the vital contributions animals make to business, tourism and rural life and it demands serious attention in policy, ethics and education (Blattner et al., 2020).

Epistemic justice

Recognising and respecting different ways of knowing, especially those that are often ignored or dismissed. In the case of animals, this means valuing how animals communicate, relate and know the world. From a donkey's hesitation to a nudge, animals express themselves in ways that challenge us to pay more attention to what they are "telling" us (Clancy et al., 2022b; Jammaers & Huopalainen, 2023).

References

Adamson, M., Kelan, E., Lewis, P., Śliwa, M., & Rumens, N. (2021). Introduction: Critically interrogating inclusion in organisations. *Organization*, 28(2), 211–227. doi:10.1177/1350508420973307.

Aguinis, H., Ashcraft, K., Benschop, Y., Blancero, D. M., Cheng, C., Cornelius, N., … & Zanoni, P. (2025). Voices from the academy: A response to President Donald Trump's anti-DEI policies. *Equality, Diversity and Inclusion: An International Journal*, 44(2), 151–157. doi:10.1108/EDI-03-2025-550.

Ahmed, A., & Rogers, M. (2017). *Working with marginalised groups: From policy to practice*. Bloomsbury. doi:10.1057/978-1-137-55957-9.

Akotirene, C. (2019). *Interseccionalidade*. Pólen Produção Editorial.

Arsel, Z., Crockett, D., & Scott, M. L. (2022). Diversity, equity and inclusion (DEI) in the *Journal of Consumer Research*: A curation and research agenda. *Journal of Consumer Research*, 48(5), 920–933. doi:10.1093/jcr/ucab057.

Blattner, C. E., Coulter, K., & Kymlicka, W. (Eds). (2020). *Animal labour: A new frontier of interspecies justice?* Oxford University Press. doi:10.1093/oso/9780198846192.001.0001.

Braidotti, R. (2019). A theoretical framework for the critical posthumanities. *Theory, Culture & Society*, 36(6), 31–61. doi:10.1177/0263276418771486.

Clancy, C., Watson, T., & Raw, Z. (2022a). Resilience and the role of equids in humanitarian crises. *Disasters*, 46(4), 1075–1097. doi:10.1111/disa.12501.

Clancy, C., Cooke, F., & Gorman, R. (2022b). Learning with donkeys: A "more-than-human" approach to animal-assisted activities. *Society & Animals*, 32(4), 341–361. doi:10.1163/15685306-bja10103.

Crenshaw, K. (1991). Mapping the margins: Intersectionality, identity politics and violence against women of color. *Stanford Law Review*, 43(6), 1241–1299. doi:10.2307/1229039.

Cudworth, E. (2014). Beyond speciesism: Intersectionality, critical sociology and the human domination of other animals. In N. Taylor & R. Twine (Eds), *The rise of critical animal studies: From the margins to the centre* (pp. 19–35). Routledge.

Donaldson, S., & Kymlicka, W. (2011). *Zoopolis: A political theory of animal rights.* Oxford University Press.

Everett, S., & Hill, I. (Eds). (2025). *Diversity, equity and inclusion for business and management*. Sage.

Ferdman, B. M. (2017). Paradoxes of inclusion: Understanding and managing the tensions of diversity and multiculturalism. *The Journal of Applied Behavioral Science*, 53(2), 235–263. doi:10.1177/0021886317702608.

Fitzsimmons, S., Özbilgin, M. F., Thomas, D. C., & Nkomo, S. (2023). Equality, diversity and inclusion in international business: A review and research agenda. *Journal of International Business Studies*, 54(8), 1402–1422. doi:10.1057/s41267-023-00642-x.

Geiger, M. (2024). Hoof work: The feminisation of donkeys in Ethiopia. *Sociological Research Online*, 29(4), 842–860. doi:10.1177/13607804231213559.

Geiger, M., Hockenhull, J., Buller, H., Engida, G. T., Kedir, M. J., Goshu, L., Getachew, M., Banejee, A., Burden, F., & Whay, H. R. (2023). Being with donkeys: Insights into the valuing and wellbeing of donkeys in Central Ethiopia. *Society & Animals*, 32(5–6), 602–630. doi:10.1163/15685306-bja10134.

Geiger, M., Hockenhull, J., Buller, H., Kedir, M. J., Engida, G. T., Getachew, M., Burden, F., & Whay, H. R. (2021). Comparison of the socio-economic value and welfare of working donkeys in rural and urban Ethiopia. *Animal Welfare*, 30(3), 269–277. doi:10.7120/09627286.30.3.004.

Hellerstedt, K., Uman, T., & Wennberg, K. (2024). Fooled by diversity? When diversity initiatives exacerbate rather than mitigate bias and inequality. *Academy of Management Perspectives*, 38(1), 23–42. doi:10.5465/amp.2021.0206.

Hribal, J. (2012). Animals are part of the working class reviewed. *Borderlands e-Journal*, 11(2), 1–37.

Iyer, A. (2022). Understanding advantaged groups' opposition to diversity, equity and inclusion (DEI) policies: The role of perceived threat. *Social and Personality Psychology Compass*, 16(5), e12666. doi:10.1111/spc3.12666.

Jackson, Z. I. (2020). *Becoming human: Matter and meaning in an antiblack world.* New York University Press.

Jammaers, E., & Huopalainen, A. (2023). "I prefer working with mares, like women, difficult in character but go the extra mile": A study of multiple inequalities in equine (sports) business. *Gender, Work & Organization*, 30(6), 2049–2068. doi:10.1111/gwao.13044.

Kim, C. J. (2015). *Dangerous crossings: Race species and nature in a multicultural age*. Cambridge University Press. doi:10.1017/CBO9781107045392.

Ko, A., & Ko, S. (2017). *Aphro-ism: Essays on pop culture, feminism and black veganism from two sisters*. Lantern.

Köllen, T. (2021). Diversity management: A critical review and agenda for the future. *Journal of Management Inquiry*, 30(3), 259–272. doi:10.1177/1056492619868025.

Kopnina, H., & Gjerris, M. (2015). Are some animals more equal than others? Animal rights and deep ecology in environmental education. *Canadian Journal of Environmental Education*, 20, 108–122.

Kossek, E. E., Vaziri, H., Perrigino, M. B., Lautsch, B. A., Pratt, B. R., & King, E. B. (2025). Reenvisioning family-supportive organisations through a diversity, equity and inclusion perspective: A review and research agenda. *Journal of Management*, 51(6), 2520–2548. doi:10.1177/01492063241310149.

Lugones, M. (2010). Toward a decolonial feminism. *Hypatia*, 25(4), 742–759. doi:10.1111/j.1527-2001.2010.01137.x.

Mkono, M. (2019). Neo-colonialism and greed: Africans' views on trophy hunting in social media. *Journal of Sustainable Tourism*, 27(5), 689–704. doi:10.1080/09669582.2019.1604719.

Mkono, M., & Hughes, K. (2025). A state-of-the-art review of animals in tourism: Key debates and future directions. *Tourism Geographies*, 27(3–4), 653–665. doi:10.1080/14616688.2024.2342462.

Monterrubio, C. (2025). Destigmatising interspecies relations: Donkeys in Mexican tourism. In C. Monterrubio, X. López-Medellín, K. Dashper & H. Wadham (Eds), *Human–animal relations in tourism, leisure and development: Perspectives from Latin America* (pp. 36–48). CABI. doi:10.1079/9781800626829.0004.

Monterrubio, C., Dashper, K., & Hernández-Espinosa, R. (2023). Horseracing as gendered leisure events: Doing and redoing masculinities in Mexico. *Leisure Studies*, 42(4), 553–567. doi:10.1080/02614367.2022.2115115.

Monterrubio, C., Dashper, K., Mendoza-Ontiveros, M. M., & Wadham, H. (2024a). Legitimizing and transforming gender relations within the contemporary equestrian sport of Charrería in Mexico. *Sociology of Sport Journal*, 42(2), 159–170. doi:10.1123/ssj.2023-0162.

Monterrubio, C., Dashper, K., & Wadham, H. (2024b). Contested intangible heritage: Equestrian sport and animal welfare in Mexico. *International Journal of Heritage Studies*, 30(3), 332–348. doi:10.1080/13527258.2023.2294761.

Mor Barak, M. E. (2022). *Managing diversity: Toward a globally inclusive workplace*. Sage.

Narayanan, Y. (2025). Animal–state relations: A critical multispecies geopolitics of animality. *Progress in Human Geography*. doi:10.1177/03091325251348611.

Ndlovu-Gatsheni, S. J. (2015). Decoloniality as the future of Africa. *History Compass*, 13(10), 485–496. doi:10.1111/hic3.12264.

Norris, S. L., Little, H. A., Ryding, J., & Raw, Z. (2021). Global donkey and mule populations: Figures and trends. *Plos One*, 16(2), e0247830. doi:10.1371/journal.pone.0247830.

Park, C. H., Park, S., & Kwon, B. (2025). Forty-five years of research on diversity, equity and inclusion in management. *Management Decision*, 63(13), 66–95. doi:10.1108/MD-11-2023-2181.

Perlo, K. (2022). What is the animal class? Idea, position, goal. *Politics and Animals*, 8, 1–13.
Riggs, D. W., Rosenberg, S., Fraser, H., & Taylor, N. (2021). *Queer entanglements: Intersections of gender, sexuality and animal companionship*. Cambridge University Press. doi:10.1017/9781108772358.
Roberson, Q. M. (2006). Disentangling the meanings of diversity and inclusion in organizations. *Group & Organization Management*, 31(2), 212–236. doi:10.1177/1059601104273064.
Rodrigues, J. B., Raw, Z., Santurtun, E., Cooke, F., & Clancy, C. (2021). Donkeys in transition: Changing use in a changing world. *Brazilian Journal of Veterinary Research and Animal Science*, 58, e174325–e174325. doi:10.11606/issn.1678-4456.bjvras.2021.174325.
Sayers, J., Hamilton, L., & Sang, K. (2019). Organizing animals: Species, gender and power at work. *Gender, Work & Organization*, 26(3). doi:10.1111/gwao.12277.
Tallberg, L., Huopalainen, A., & García-Rosell, J. C. (2024). Beyond anthropocentrism: A call to action for multispecies inclusivity. *Business & Society*. doi:10.1177/00076503241271254.
Taylor, N., & Sutton, Z. (2018). For an emancipatory animal sociology. *Journal of Sociology*, 54(4), 467–487. doi:10.1177/1440783318815335.
Tomlinson, M. (2024). "Clean communication": Felt-sense methodologies and the reflexive researcher in equine-assisted personal development. *The Sociological Review*, 72(1), 155–174. doi:10.1177/00380261231186752.
Triana, M. D. C., Kim, K., Byun, S. Y., Delgado, D. M., & Arthur, W.Jr (2021). The relationship between team deep-level diversity and team performance: A meta-analysis of the main effect, moderators and mediating mechanisms. *Journal of Management Studies*, 58(8), 2137–2179. doi:10.1111/joms.12670.
Van Bommel, H. M., Hubers, F., & Maas, K. E. H. (2024). Prominent themes and blind spots in diversity and inclusion literature: A bibliometric analysis. *Journal of Business Ethics*, 192(3), 487–499. doi:10.1007/s10551-023-05522-w.
Wadham, H., & Dashper, K. (2024). The slow road to sustainability: An interspecies perspective on decent work. *Journal of Sustainable Tourism*, 33(9), 1816–1831. doi:10.1080/09669582.2024.2374361.
Watson, T., & Clancy, C. (2025). Shared work? Unravelling interspecies entanglements, agency and the rhythms of equids at work. *Frontiers in Veterinary Science*, 12, 1570879. doi:10.3389/fvets.2025.1570879.
Watson, T., Kubasiewicz, L. M., Chamberlain, N., Nye, C., Raw, Z., & Burden, F. A. (2020). Cultural "blind spots," social influence and the welfare of working donkeys in brick kilns in northern India. *Frontiers in Veterinary Science*, 7, 214. doi:10.3389/fvets.2020.00214.
Watson, T., Kubasiewicz, L. M., Nye, C., Thapa, S., Norris, S. L., Chamberlain, N., & Burden, F. A. (2022). "Not all who wander are lost": The life transitions and associated welfare of pack mules walking the trails in the mountainous Gorkha region, Nepal. *Animals*, 12(22), 3152. doi:10.3390/ani12223152.
Waverley, J. (2024). Organs or bodies? Toward an equitable, embodied and animal-inclusive diversity, equity and inclusion agenda. *Consumption Markets & Culture*, 27(2), 233–250. doi:10.1080/10253866.2023.2276419.

PART IV

Animals, governance and institutional frames

10

ANIMALS AND ACCOUNTING AND VALUE

Introduction

Accounting and value are often presented as neutral ideas – ways of measuring what things are worth, allocating resources and reporting performance (Granà et al., 2025). Yet when applied to nonhuman animals, these ideas raise deep ethical and political questions. Accounting does not simply record the world; it helps to make it. It shapes what – and who – becomes visible (Vosselman, 2022).

In capitalist economies, animals usually appear in accounting systems in limited and fragmented ways. As commodities, their bodies and products are priced and sold for profit (e.g. meat, milk or leather). As resources, they are treated as renewable assets that generate income. And as externalities, their suffering, deaths or ecological damage are recorded only as side-effects rather than moral concerns (Himick & Vinnari, 2023). Animals thus occupy what Wadiwel (2023) calls a "hybrid value position" within capitalism – they are at once raw materials, labour and products. Their bodies produce value through metabolic work while also becoming commodities exchanged on the market. This overlap between worker and product exposes how accounting helps to disguise dependence on animal life as neutral calculation.

Drawing on the critical work of Dinesh Wadiwel, this chapter examines how these ways of measuring value have developed and how they might be transformed. We argue that accounting sits at the crossroads of economic rationality, ethical invisibility and systemic violence. An interspecies perspective – one that recognises animals as participants in, not just resources for, business and society – is therefore essential for reimagining what counts as value. In response, this chapter contributes to the book's broader project of developing an interspecies approach to responsible management by

DOI: 10.4324/9781003585541-15

showing how accounting – one of business's core decision-making tools – shapes whose lives are recognised and whose are erased.

The value of a more animal-centred approach

Accounting is never purely technical. It is also moral and political because it decides what matters. Vinnari & Vinnari (2022) show that animals are largely invisible in sustainability accounting because they fall awkwardly between the categories of "society" and "nature." Domesticated animals are recorded as part of the economy, while wild animals are relegated to "the environment." Neither are recognised as subjects with their own interests.

This invisibility is not accidental. As Dinesh Wadiwel (2023) explains, it stems from what he colourfully calls the "handshake between anthropocentrism and capitalism." By this he means the long-standing alliance between a worldview that places humans at the centre and an economic system driven by endless growth and profit. This handshake legitimises domination over other species while presenting it as rational and inevitable. Accounting sustains the handshake by turning animal life into financial data – counting bodies, costs and outputs while leaving ethical questions outside the frame.

Taking a more animal-centred approach to accounting matters because it reveals the violence hidden in these systems of measurement. Cuckston (2018) argues that even progressive forms of "biodiversity accounting" often remain anthropocentric: They value species only when they provide benefits to humans or reduce business risk. Similarly, Vinnari and Vinnari (2022) show how sustainability metrics erase individuality by grouping living beings together as "natural capital." For Wadiwel, this translation of life into data is itself a form of domination. In capitalist systems, animals become what he calls "value-producing bodies." Their physical and reproductive capacities – their growth, milk, eggs or flesh – are used to generate profit. The moral cost of this process is disguised by the language of efficiency, productivity and sustainability (Milne & Gray, 2013).

To think differently about accounting, we need to recognise how it turns lives into numbers. Traditional accounting depends on comparing everything through a single measure of value, usually money. This logic hides the fact that living beings are not equivalent or interchangeable (Wadiwel, 2023). Gunderson (2013) traces how the rise of capitalism replaced animals' "use value," or their role in meeting human needs, with "exchange value," or market price, transforming them from companions or co-workers into commodities. Clark and Wilson (2020) describe this as a historical rupture that alienated humans from nature and produced what they call "alienated speciesism." Both studies show that the commodification of animals is not a by-product of industry but a structural feature of capitalism itself. Accounting, as the language of that system, plays a central role in concealing the ethical and ecological costs of profit.

An animal-centred approach challenges this picture. If we recognise animals as labouring subjects and ethical beings – participants in, not merely products of, the economy – we can rethink what counts as value. Vinnari and Vinnari's (2022) proposal to "make the invisibles visible" advances this goal by developing indicators that measure progress from exploitation to rights and care. Yet, as Wadiwel warns, inclusion alone is not enough. Unless we question the wider logic of growth, measurement and profit, we risk reproducing the same structures of domination in a more palatable form.

Introducing our expert collaborator

To explore these issues further, this chapter draws on the work of Australian political theorist and critical animal-studies scholar Dinesh Wadiwel, whose research connects Marxist political economy, posthumanist ethics and animal-rights theory. Across *The War Against Animals* (2015) and *Animals and Capital* (2023), Dinesh argues that human–animal relations are organised around a continuous "war": A system of domination that is both economic and moral. In our interview, he explained that capitalism and anthropocentrism reinforce one another – capitalism demands endless growth, while anthropocentrism provides moral justification for exploiting animals to achieve it. Accounting, as the key mechanism for making value visible, therefore acts as the medium through which this partnership operates.

In Dinesh's analysis, animals hold a hybrid position within capitalism: They are at once raw materials, commodities and labour. Farmed animals, for example, both embody and create value. They are products to be sold but also living machines whose bodily activity – eating, reproducing, growing – generates surplus value. Their lives and deaths occur at the meeting point of use and exchange, nature and labour. The factory farm, Dinesh argues, is the clearest expression of this system, or the handshake between anthropocentrism and capitalism "made flesh" (Wadiwel, 2023). Accounting practices such as feed-conversion ratios, performance metrics or welfare audits translate animal exploitation into neutral numbers. These metrics make the system appear efficient and humane while masking the suffering and coercion that sustain it.

Dinesh's work builds on and extends the ideas of Karl Marx, feminist scholars and posthumanist thinkers. While Marx rarely focused on animals directly, he showed how capitalism turns both humans and animals into commodities by separating them from any intrinsic worth (1867/1976). Feminist scholars such as Carol Adams (1990) and Donna Haraway (2008) highlight how reproduction and care – activities often ignored in economics – are central to exploitation across species. Finally, posthumanist scholars challenge the human-centred worldview altogether, suggesting that agency and value are distributed across human and non-human life. Bringing these insights together, Dinesh shows that animals are not passive victims but central to how capitalism produces wealth. Recognising this interdependence forces us to see that our economies rely on multispecies labour – and that ignoring this fact leaves ethical and sustainability debates incomplete.

The rest of the chapter develops these ideas in four parts. The conceptual foundations section reviews how accounting and sustainability research have treated animals as resources or risks. The next section develops an interspecies approach to accounting and value, by expanding on Dinesh's concept of animals' hybrid value positions – as raw material, commodity and labour – and exploring how accounting hides violence through the language of efficiency. Later sections consider the ethical and practical implications of this analysis, showing how rethinking accounting could inform responsible management, multispecies justice and new ways of defining value. Finally, the chapter turns to teaching and learning, suggesting how an interspecies perspective can help future managers and accountants see accounting not only as a technical process but as a moral and political act.

The conceptual foundations of our ideas about accounting and value

Accounting has deep historical roots. It began as a practical way to track exchanges, debts and possessions – an early technology for recording relationships between people and things (Miller & Power, 2013). Anthropologists tell us these systems were never purely economic: In many "traditional" societies, practices of gift, reciprocity and redistribution expressed social ties and moral duties as much as trade (Sahlins, 1972/2013). Recent historical research further suggests that environmental and ecological accounts predated financial accounting, documenting species, land and resource relations long before the emergence of modern commercial calculation (Atkins et al., 2023). Early forms of record-keeping registered bonds not only among people, but between humans and the more-than-human world. Over time, this moral, relational dimension gave way to more instrumental, abstract calculation tied to commerce and then capitalism. As markets expanded, accounting evolved from a social practice of reciprocity into a system for measuring performance and managing risk, often erasing earlier moral or ecological concerns (De Villiers et al., 2020). It became the foundation of modern economic life, allowing wealth to be measured, compared and accumulated across time and place (Vollmer, 2021).

Because of this history, accounting is closely linked to particular ways of understanding value. Its methods are grounded in human perspectives – what matters to people, organisations and markets. Even in its most sophisticated forms, accounting measures the world through human categories of ownership, utility and exchange (Bebbington & Gray, 2001; Wolfe, 2003). Animals, landscapes and ecosystems appear only when they affect these human-defined flows of value. In this sense, accounting is built on an anthropocentric worldview, one that assumes the economy exists for human benefit and that nonhuman life is valuable only when it contributes to human goals (Pesci et al., 2024).

Sustainability and the illusion of balance

Malm (2016) and others situate this crisis within the consolidation of fossil capitalism, where calculative logics and extraction became mutually reinforcing. The anthropocentric worldview outlined above thus emerges also within sustainability and environmental accounting. Here, the natural world is often reframed as "capital" – something that can be invested in, depleted or replenished. Forests, rivers and animal populations are described as "assets," while pollution, extinction or cruelty become "costs" or "externalities." These terms give environmental issues a place within business reporting, but they also reshape them to fit economic logic. As Tavakolnia (2023) points out, this kind of accounting does not prevent harm: it simply translates it into the language of finance. Once something can be priced, it can also be justified or traded off. Atkins and Maroun (2018) similarly argue that extinction accounting struggles to challenge biodiversity loss when it remains tied to the same calculative frameworks that structure corporate decision-making. Framing extinction as a matter of disclosure or capital management risks reinforcing, rather than challenging, the anthropocentric assumptions that make species loss thinkable in the first place.

Milne and Gray's (2013) influential critique of sustainability reporting shows how this translation process hides deeper moral questions. In their view, frameworks like the "triple bottom line" or "natural capital accounting" present ecological destruction as a problem of measurement rather than of ethics or power. Yusoff (2018) reminds us that such extractive logics are historically and materially entangled with racialised regimes of valuation and disposability. By using the language of balance, efficiency and productivity, such frameworks make ongoing exploitation appear rational and responsible. Drury et al. (2023) extend this argument to the wider field of sustainability, showing that its managerial focus on performance and control often reproduces the same logic of domination that it claims to challenge.

Within accounting research, these critiques have helped to form what is now called "critical accounting," an approach that examines how accounting practices shape social and environmental realities. Instead of treating accounting as a neutral tool, critical scholars see it as a cultural and political practice that reflects the values of the society that created it (Gendron, 2018). From this perspective, the problem is not simply that animals or ecosystems are missing from the books, but that the very idea of "value" in accounting is defined in human and economic terms.

Several scholars have traced how this human-centred idea of value developed. Vosselman (2022) highlights that accounting's claim to objectivity – its promise to represent the world as it is – rests on a narrow, managerial way of seeing. It focuses on what can be counted and compared, ignoring relationships and experiences that do not fit numerical form. Deplazes-Zemp (2024) reaches a similar conclusion from a moral philosophy standpoint, arguing that the old divide between *intrinsic* and *instrumental* value is too limited because it treats humans as the ones who decide what is worth protecting. From this angle, accounting becomes not just a mirror of the economy but a mechanism that continually recreates human dominance (Maroun et al., 2023).

The emergence of more critical understandings

These critical writers have thus begun to ask how these assumptions might be unsettled. Himick and Vinnari (2023) interrogate the anthropocentric foundations of accounting, asking whether existing frameworks can accommodate nonhuman stakeholders without reproducing extractive assumptions. Vollmer (2021) imagines a form of "planetary accounting" that would recognise nonhuman beings as part of the public community of concern. Similarly, McLaren and Appleyard (2022) explore how corporate reporting on animal welfare exposes the tensions between moral care and market logic. Folkers (2024) contributes from a sociological angle, showing how climate accounting tries to connect environmental futures with financial ones, often reconfiguring ecological concerns through the logics of valuation and investment. Across these different perspectives, a shared insight emerges: Accounting both reflects and reinforces the power structures of capitalism and its anthropocentrism is a central part of that process.

For critical scholars, then, the challenge is not only to include animals or nature in existing accounting systems but to question how those systems define value in the first place (Hamilton, 2017). As Tavakolnia (2023) and Vollmer (2021) suggest, real change requires moving beyond the idea that everything must be measured financially to matter. The task is to recognise that accounting is a moral and political practice, one that has long served human and economic priorities while concealing their consequences. Understanding this history sets the stage for the next part of the chapter, which explores how emerging interspecies perspectives are rethinking what it means to account for value in a shared, more-than-human world.

Towards an interspecies approach to accounting and value

The previous section examined the conceptual foundations of accounting and showed how it emerged from human-centred systems of exchange and measurement. We noted how accounting's reliance on monetary equivalence made it both a moral and political technology – one that privileges human interests and reduces other forms of life to resources or "natural capital." This section moves the discussion forward by asking how we might rethink accounting and value beyond the human.

To do so, we draw on the work of Dinesh Wadiwel – particularly his books *Animals and Capital* (2023) and *The War Against Animals* (2015) – which provide a framework for analysing how animal life is captured within capitalist systems of value creation. Dinesh's argument connects directly to the concerns raised by scholars such as Cuckston (2018), Milne and Gray (2013) and Vinnari and Vinnari (2022), who have shown how accounting translates ethical questions into technical problems of measurement. Dinesh extends this critique by revealing how capitalism depends not only on human labour but also on the ongoing appropriation of animal life.

As outlined in the introduction, animals hold three overlapping value positions within capitalist economies – as raw materials, commodities and labour (Wadiwel, 2023). This section develops that idea in more depth, showing how these categories interact and reveal the moral and material contradictions of accounting. By tracing how animals move between these positions, we can better understand the systems that render them simultaneously visible as assets and invisible as living beings. Building on this, the discussion opens the way for an interspecies approach to value – one that recognises animals as participants, not just products, in economic life.

Animals' hybrid positions in capitalism

Dinesh begins from a deceptively simple observation: Animals are everywhere in the global economy. They are farmed, traded, consumed and experimented on; they are entertainment, symbols and capital. Yet within economic and accounting frameworks, their presence is almost invisible. When they do appear, it is as stock, property or externality.

To make sense of this paradox, Wadiwel (2023) identifies three interconnected ways that capitalism positions animals. As raw materials, animals provide the biological inputs of production: Their bodies, flesh, eggs, milk, skin and genes form the basis of entire industries like agriculture, pharmaceuticals and fashion. They are treated as renewable resources to be extracted, optimised and replaced. As commodities, animals and their products circulate as saleable goods. Once transformed into meat, milk, leather or pet food, their individuality is erased – replaced by a barcode and a price.

The third position, labour, is less visible but no less important. Here, Dinesh builds on Marx's concept of labour as the process that creates value. For animals, particularly those used in food systems, labour takes the form of metabolism, growth and reproduction. The chicken that turns feed into muscle, or the cow that converts grass into milk, performs what Barua (2025) calls "metabolic work." This is not metaphorical: It is labour that produces surplus value for capital. Animals' very bodies become sites of production – their flesh both the means and the outcome of work.

These overlapping roles create what Dinesh describes as a hybrid position. In animal industries, the worker and the commodity are one and the same. The animal's living body produces value and simultaneously becomes the product that is sold. This blurring of producer and product, as Clark and Wilson (2020) and Gunderson (2013) note, represents a historical rupture in human–animal relations. In pre-capitalist contexts, animals were co-workers in agriculture and transport. Under industrial capitalism, they became units of extraction. Fraser (2014) thus describes capitalism as dependent on such "hidden abodes" of reproduction and care – domains essential to accumulation yet systematically obscured.

Accounting practices both reflect and reinforce this transformation. Animals are entered into ledgers as "biological assets," their lives valued at fair market price. Costs of feed, veterinary care and housing are calculated against the expected sale value of their bodies. Nowhere in this equation is there recognition of animals as sentient subjects. Himick and Vinnari (2023) describe this as ethical invisibility – the exclusion of animal interests from the very mechanisms that manage their exploitation. Vinnari and Vinnari (2022) note that animals occupy an ambiguous position in accounting – neither fully part of "society" nor "nature." Yet this ambiguity is not a feature of animals themselves but a limitation of accounting's anthropocentric design, which fails to recognise beings that do not fit neatly into its human-centred categories. For Wadiwel, this invisibility is not accidental but systemic. This is a "war" against animals – continuous, institutionalised violence that underpins capitalist prosperity. Animals' hybrid positions reveal capitalism's dependency on life that is simultaneously productive and expendable. For managers and accountants, this is not abstract theory. It shapes how costs are calculated, how efficiency is defined and how responsibility is framed within everyday organisational practice.

The factory farm: Where labour meets capital

Nowhere is this dependency clearer than in the factory farm. Here, the logic of capital – endless accumulation and efficiency – meets the biological reality of animal life (see also Chapter 2 and Chapter 4). Within this space, every process is designed to accelerate production. Animals' bodies are engineered through selective breeding, controlled feeding and environmental manipulation to achieve faster growth and higher yield. Chickens reach slaughter weight in six weeks; dairy cows produce ten times more milk than they would naturally. These changes, celebrated in industry reports as progress, are recorded in accounting systems as improvements in feed-conversion ratios, mortality rates and output per animal. As Milne and Gray (2013) argue, this transformation represents accounting's ability to disguise moral questions as technical ones. What appears as an "efficiency gain" is in fact an intensification of exploitation. Cuckston (2018) extends this critique to biodiversity accounting, noting that even environmental reforms often remain anthropocentric, valuing species only for their utility to humans.

Corporate welfare schemes reinforce this logic by making ethics measurable. McLaren and Appleyard (2022) analyse the Business Benchmark on Farm Animal Welfare (BBFAW) to show how companies turn moral responsibility into scores and rankings. Firms that disclose more information receive higher ratings, regardless of actual welfare outcomes. Similarly, Gottlieb et al. (2022) find that welfare audits shape farmers' sense of accountability – they become responsible for data points such as mortality, weight gain and productivity rather than for the lived wellbeing of the animals in their care.

Wadiwel interprets such mechanisms as technologies of domination. The more completely animals are measured, the more thoroughly they are controlled. Accounting systems reduce life to what can be counted, transforming suffering into statistics. In this way, the factory farm functions not only as a physical site of exploitation but also as a symbolic one: it embodies capitalism's capacity to turn life itself into a commodity. This analysis also resonates with feminist scholars like Adams (1990) and Haraway (2008), who have long shown how reproduction, care and bodily labour are central yet devalued in capitalist systems. Dinesh's inclusion of animal labour extends this insight across species, exposing the continuities between the exploitation of human and nonhuman bodies.

Technologies of domination: Making animals calculable

Beyond the farm, the logic of domination extends through what Foucault might call a machinery of governmentality. Accounting tools such as sustainability audits, welfare certifications and carbon assessments make animals calculable yet invisible. Each device claims to promote responsibility, but in practice it translates ethics into management.

As Tavakolnia (2023) observes, financial valuation can turn moral problems into trade-offs: Once suffering is priced, it becomes something that can be optimised or offset. Drury et al. (2023) thus show that sustainability reporting, with its emphasis on performance and control, often reproduces the same power relations it aims to reform.

In response, critical accounting researchers have developed the concept of counter-accounts. Himick and Vinnari (2023) define these as alternative forms of accounting – reports, documentaries or artistic works – that challenge dominant narratives. Examples include NGO exposés of slaughterhouse conditions or visual campaigns showing the reality of intensive farming. Vinnari and Laine (2017) describe counter-accounts as moral mechanisms that replace abstraction with emotion, encouraging empathy and ethical reflection. However, Dinesh (Wadiwel, 2015, 2023) cautions against assuming that inclusion or visibility automatically leads to justice. Audits and counter-accounts can be absorbed back into the system, becoming what he calls "technologies of containment." They manage moral discomfort without changing the underlying logic of exploitation. For example, improved welfare standards may make consumers feel ethical while leaving production volumes – and deaths – unchanged. Dillard and Vinnari (2019) therefore warn that real reform needs to address the underlying values and assumptions of accounting, not just increase what it measures or calculates.

Rethinking value: Encounter, resistance and counter-practice

If capitalist accounting defines value through profit and control, an inter-species approach calls for a radical redefinition. Barua (2016) proposes moving from exchange value to what might be called encounter value – the moral and relational worth that arises in meetings between species. This form of value cannot be reduced to price because it depends on recognition, care and vulnerability. Philosophers such as Deplazes-Zemp (2024) support this shift, arguing that the old divide between intrinsic and instrumental value keeps humans at the centre of moral reasoning. Encounter value instead foregrounds relation: Worth emerges through interaction, not hierarchy. Within accounting, this could mean designing systems that recognise shared dependencies rather than one-sided exploitation.

Celermajer et al. (2023) develop a related idea of multispecies justice, urging institutions to consider nonhuman beings as members of the moral community. In practice, this might involve expanding environmental, social and governance (ESG) metrics to include animal wellbeing or developing new forms of multispecies accounting (Vinnari & Vinnari, 2022). Such systems would not simply quantify welfare but reflect on the moral and ecological consequences of production. Some scholars connect this rethinking of value to degrowth economics, which advocates reducing material throughput and prioritising wellbeing over profit. Within accounting, this could mean evaluating success through indicators of care, time or flourishing rather than yield. Counter-accounts by activists and artists – such as photo essays documenting animal lives – already point towards this possibility. They make suffering visible not as data but as a call for empathy, echoing what Himick and Vinnari (2023) describe as a shift from calculation to compassion. Encounter value and multispecies accounting offer a route to repairing this alienation by reintroducing relationality into how we measure and understand worth.

In summary, Dinesh's work demonstrates that animals are not peripheral to capitalism – they are foundational. Their bodies, movements and reproductive capacities generate immense value, yet accounting systems render this value production visible only in the form of money value. In this respect, accounting acts as both a moral veil and an engine of exploitation. By tracing animals' hybrid value positions, examining the factory farm as the meeting point of labour and capital and analysing the technologies that make animals calculable, we see how deeply anthropocentric logics structure modern economies. But by exploring ideas such as encounter value, counter-accounts and multispecies justice, we also glimpse alternative ways of thinking. Critical accounting scholars – from Milne and Gray's (2013) critique of sustainability rhetoric to Vinnari and Vinnari's (2022) call to "make the invisibles visible" – share Wadiwel's aim of unsettling capitalism's moral foundations. Together, they suggest that accounting can no longer be confined to human interests. It must grapple with the fact that its ledgers contain lives. An interspecies approach therefore asks us to account otherwise – to see value not just in profit or productivity but in coexistence, care and resistance. The goal is not to make capitalism slightly more ethical but to challenge the very assumptions that make domination appear rational (Wadiwel, 2023). Accounting, reimagined in this way, becomes a practice of moral imagination – a way to see and respond to the shared vulnerabilities that define life on a more-than-human planet.

Directions for future research

Future research needs to move beyond consumer ethics – ideas about what individuals should or should not buy – and towards what Wadiwel calls "*production politics*." This means paying attention to how food and production systems are organised in the first place. For more than 50 years, the poor conditions of factory farming have been well known, yet consumer habits have changed little. Wadiwel argues that the problem is not a lack of moral awareness but the fact that the economic system itself depends on the continual use of animals. Even when consumers buy "ethical" products, the wider structures of production remain intact. Researchers can also fall into this trap when they study animal suffering as a moral issue to be corrected, rather than as a central feature of how the economy works. Dinesh urges scholars not to stand outside production looking in, but to examine how systems that "create" value – farming, logistics, accounting – make exploitation appear normal, efficient and profitable. Shifting from consumption to production is therefore not only a political move but a research challenge. It asks us to study the mechanisms that make animal use seem inevitable and to imagine how they could be otherwise.

From this perspective, accounting research has an important role to play, but also clear limits. Rather than refining systems that treat animal life as cost, output, or risk, scholars could develop approaches that make visible the labour and suffering through which value is produced. The idea of *metabolic work* offers a useful starting point (Barua, 2025; Wadiwel, 2023). If animals' growth, reproduction and bodily activity contribute to surplus value, accounting might be rethought to recognise these processes as labour rather than as automatic or "natural" inputs.

At the same time, Marx's *Capital* makes clear that capitalist accounting systems do not fully record surplus value, whether produced by human or non-human labour. As Bryer (2017) shows, Marx developed his theory of value in close engagement with accounting practices precisely to explain how surplus value is calculated and controlled while remaining partially invisible. From this perspective, *counter-accounting* should be understood not as a technical adjustment, but as a radical alternative to dominant accounting logics: one that challenges how value itself is measured, distributed and obscured across species.

Seen this way, future research should not aim simply to "add" animals into existing ledgers. Instead, it could ask how accounting might acknowledge animals' contribution to production without reducing them to capital. For example, what changes when a cow's milk or a hen's egg is treated as embodied labour rather than yield? Researchers might also explore how death, exhaustion and ecological damage could be recognised as real losses rather than neutral production costs. Alongside this, creative counter-accounts – ethnographic, visual or artistic – could experiment with ways of representing animal labour and interdependence that resist being folded back into extractive metrics.

Advancing this agenda will depend on interdisciplinary collaboration. Accounting, food-systems and critical animal-studies scholars could work together to examine how value and power circulate across multispecies economies. Such work could reveal how animal life, labour and environments are connected through global supply chains and how accounting might represent these interdependencies more ethically. Insights from geography, systems theory and ecological economics could support the design of "*post-growth*" frameworks that measure wellbeing, care and sustainability rather than profit alone. Dinesh's call for new political imaginaries invites economists, futurologists, sustainability researchers, ecologists and more to imagine economies that recognise animal labour and ecological limits as central to production. By combining technical innovation with moral imagination, future research can move beyond describing exploitation to rethinking the systems of measurement that sustain it – developing tools for economies truly fit for all species.

Ethical implications

The ideas explored across this chapter have important ethical implications for how we understand accounting and value. Wadiwel's analysis asks us to see accounting not as a neutral language but as a practice that sustains violence by erasing animal subjectivity. When animals appear as "biological assets," productivity ratios or welfare scores, their lives are translated into data points that can be optimised. This process does moral work: it normalises killing and exploitation by making them seem necessary, efficient and even responsible. *The War Against Animals* and *Animals and Capital*, argues that capitalism severs "intrinsic value" from "use value." Under this logic, beings matter only insofar as they can be used and accounting makes that use calculable and manageable. The cow's lactation or the hen's laying becomes "metabolic work" to be costed. Death becomes an input-output problem and suffering a variance to be corrected.

A more ethical approach to accounting would begin by questioning how systems of calculation make exploitation appear acceptable. It would ask what is lost when living beings are reduced to prices or performance figures and whether some forms of harm simply cannot be measured. Instead of adding more data or indicators, the challenge is to rethink what we choose to count. This means recognising animals as subjects whose work, suffering and needs cannot be captured by metrics of output or efficiency. Ethical accounting should therefore draw a clear line between what can be priced and what should never be.

Practical implications for responsible management

The discussion so far has shown that accounting is never neutral: It shapes what – and who – is recognised as valuable. If we take seriously Dinesh's argument that capitalism depends on a continuous war against animals, then responsible management cannot simply mean improving existing practices. It must involve rethinking what responsibility itself looks like within systems built on exploitation. This section explores what that might mean for managers and leaders – individuals who, while working within the machinery of capitalism, still have opportunities to act differently and open cracks in the system.

For individual managers, these ideas can feel both important and daunting, especially where they are operating within industries that rely on animal life. In farming, fashion, research or logistics, for example, these systems are not easily changed. Capitalism's war against animals is not a set of isolated wrongs but part of how value is created (Wadiwel, 2015). No single leader can end this structure. Still, managers are not powerless. Responsibility can begin in small, everyday acts of reflection and resistance – finding and working within the fissures of capitalism.

Rethinking value creation in multispecies terms means starting where one stands. Managers can ask how animals contribute to value in their organisation – not as abstract "inputs" or "biological assets," but as living beings whose bodies, labour and environments make production possible. Making these interconnections visible can itself be an ethical act. For instance, an operations manager might trace where animal life enters a supply chain and what forms of suffering or ecological harm are involved. An accountant might treat losses usually deemed "external" – death, waste or exhaustion – as genuine, material outcomes. These small recognitions challenge the "moral invisibility" of exploitation by revealing the life behind the ledger (Wadiwel, 2015).

Responsible managers must also navigate the contradictions of green capitalism, where life is priced and managed through ESG scores, welfare audits and sustainability metrics. Dinesh warns that such systems often reproduce domination under the guise of care, turning living beings into data. Responsible management therefore involves using these tools critically – recognising the communicative value of sustainability reports for example, but refusing to confuse measurement with morality. Drawing on Deplazes-Zemp's (2024) idea of *"relational value,"* managers might instead focus on relationships between species, emphasising coexistence, care and mutual dependence rather than efficiency or profit.

Institutional change can grow from such micro-political acts. Managers can promote animal-inclusive reporting, documenting ethical and ecological impacts as part of organisational performance. They can support radical transparency in supply chains, ensuring animal lives and deaths are not hidden behind technical language. Some may work with NGOs or artists to create counter-accounts (Himick & Vinnari, 2023) that make animal experiences visible in ways numbers cannot. These efforts will not end capitalism's war against animals, but they matter. Responsible managers can use their influence to question dominant assumptions of efficiency and growth, cultivating moral imagination within the system. This chapter has repositioned accounting not as a neutral reporting tool, but as a moral practice that shapes whose lives count in business and whose do not: Even inside the machinery of profit, there remain opportunities – however modest – to account otherwise.

Reflective activities for inside and outside the classroom

A Revision questions that test knowledge ("what?")

- A1 According to the chapter, how do animals occupy "hybrid value positions" (raw material, commodity, labour) and why does this matter for accounting?
- A2 What does Dinesh mean by capitalism's "war against animals," and how do accounting practices (e.g. biological assets, feed-conversion ratios, welfare audits) help organise that war?
- A3 What is "metabolic work," and how does it challenge the idea that animals are merely "inputs" or "stocks"?
- A4 What are "counter-accounts" and how do they differ from conventional ESG or welfare reporting?

B Reflective questions that build understanding ("so what?")

- B1 If accounting shapes reality, not just records it, how might changing what appears in a company's KPIs or dashboards affect real-world decisions, e.g. how many animals are kept, how they live, or how profit is defined?
- B2 In what ways does economic rationality leave out suffering, care and relationality – and what kinds of harm become normalised as a result?
- B3 If we counted animals' labour and losses as part of production, how would that change what "productivity" or "profit" mean in industries like dairy or poultry?
- B4 When do tools that make things "visible" – like audits, ratings or welfare reports – actually help animals and when do they risk reinforcing the same systems of control they claim to challenge?

C Take action ("do")

- C1 Pick a product you buy weekly. Sketch its supply chain and mark every point where animal life, labour or death is involved. Identify what the company makes visible – and what remains off the books.
- C2 Rewrite a one-page performance report for a firm in food, fashion or pharma to include "animal-inclusive" metrics (e.g. numbers of animals involved, duration/conditions of confinement, mortality/exhaustion as losses). Note what cannot or should not be priced.
- C3 Create a short counter-account (poster, photo essay, 60-second video or infographic) which communicates realities that a standard ESG summary obscures. Aim to "inform without sanitising."
- C4 Draft a "radical transparency" pledge for an organisation (or university procurement): What specific information about animals used/affected would be published, how often and in what accessible form?

D Ideas to discuss with colleagues, friends and family ("share")

- D1 Have you ever read a company's sustainability or animal welfare statement, e.g. from a food, cosmetics or clothing brand? With a friend or family member, look at one together. Whose lives and losses are shown and whose are missing? Why do you think that is?
- D2 Have you ever seen a documentary, advert or social media clip that used numbers or data to prove a company's "success" or "responsibility"? Talk about how persuasive it felt – and what, if anything, you now think might have been left out of that story.
- D3 Think about an everyday purchase in your household, e.g. milk, pet food, makeup, shoes. Discuss with others whether it could be replaced

with a non-animal alternative. What arguments or habits make such changes feel easy or hard?

- D4 Ask friends or family what they think "responsible business" means. Then share what you've learned in this chapter, namely that accounting and value are not neutral and that caring can be as important as counting. How do they respond?

Key ideas

Hybrid value positions

In capitalist economies, animals often hold three overlapping roles at once: *raw materials* (their bodies, milk, eggs or skin), *commodities* (products to be sold) and *labour* (their living activity that creates value). This mix of roles blurs the line between worker and product, letting accounting treat them as assets while ignoring their lives and suffering (Wadiwel, 2023).

Metabolic work

When animals eat, grow, reproduce or produce milk and eggs, their bodies are doing what Wadiwel calls "metabolic work." This work creates profit, yet accounting treats it as a natural process rather than labour. Seeing it as work exposes the real costs – fatigue, injury and death – that usually go uncounted (Barua, 2025; Wadiwel, 2020).

Multispecies materiality

In accounting, "materiality" means deciding what counts as important enough to report. A multispecies approach says animal lives, deaths and wellbeing matter just as much as financial results or business risks. It pushes companies to face the full ethical and ecological consequences of their actions (Vinnari & Vinnari, 2022).

The externalities trick

Labelling animal suffering or environmental damage as "externalities" allows businesses to ignore them. These costs are real, but pushed outside the ledger. Critical accounting challenges this by arguing they should be recorded as central to performance, not hidden side effects (Milne & Gray, 2013).

Counter-accounts

Counter-accounts use stories, images or alternative reports to show what normal accounting leaves out. They reveal the lives, losses and relationships hidden behind statistics, turning ethical discomfort into public awareness and moral debate (Himick & Vinnari, 2023).

References

Adams, C. J. (1990). *The sexual politics of meat*. Continuum.

Atkins, J., & Maroun, W. (2018). Integrated extinction accounting and accountability: building an ark. *Accounting, Auditing & Accountability Journal*, 31(3), 750–786. doi:10.1108/AAAJ-06-2017-2957.

Atkins, J. F., Doni, F., McBride, K., & Napier, C. (2023). Exploring the historical roots of environmental and ecological accounting from the dawn of human consciousness. *Accounting, Auditing & Accountability Journal*, 36(6), 1473–1502. doi:10.1108/AAAJ-03-2023-6354.

Barua, M. (2016). Lively commodities and encounter value. *Environment and Planning D: Society and Space*, 34, 725–744. doi:10.1177/0263775815626420.

Barua, M. (2025) Metabolic geographies: Work, shifts and politics. *Progress in Human Geography*, 49(2), 145–163.

Bebbington, J., & Gray, R. (2001). An account of sustainability: Failure, success and a reconceptualization. *Critical Perspectives on Accounting*, 12(5), 557–587. doi:10.1006/cpac.2000.0450.

Bryer, R. (2017). *Accounting for value in Marx's Capital: The invisible hand*. Bloomsbury. doi:10.5040/9798216459217.

Celermajer, D., Schlosberg, D., Wadiwel, D., & Winter, C. (2023). A political theory for a multispecies, climate-challenged world: 2050. *Political Theory*, 51(1), 39–53. doi:10.1177/00905917221128833.

Clark, B., & Wilson, T. D. (Eds). (2020). *The capitalist commodification of animals*. Emerald.

Cuckston, T. (2018). Making accounting for biodiversity research a force for conservation. *Social and Environmental Accountability Journal*, 38(3), 218–226. doi:10.1080/0969160X.2018.1516.

De Villiers, C., Hsiao, P. C. K., & Maroun, W. (Eds). (2020). *The Routledge handbook of integrated reporting*. Routledge.

Deplazes-Zemp, A. (2024). Beyond intrinsic and instrumental: Third-category value in environmental ethics. *Ethics, Policy & Environment*, 27(2), 166–188. doi:10.1080/21550085.2023.2166341.

Dillard, J., & Vinnari, E. (2019). Critical dialogic accountability: From public to publicness. *Accounting, Auditing & Accountability Journal*, 32(1), 6–28. doi:10.1016/j.cpa.2018.10.003.

Drury, M., Fuller, J., & Hoeks, J. (2023). Embedding animals within a definition of sustainability. *Sustainability Science*, 18(4), 1925–1938. doi:10.1007/s11625-023-01310-7.

Folkers, A. (2024). Risking carbon capital: Reporting infrastructures and the making of financial climate risks. *Economy and Society*, 53(3), 504–526. doi:10.1080/03085147.2024.2348316.

Fraser, N. (2014). Behind Marx's hidden abode. *New Left Review*, 86, 55–72.

Gendron, Y. (2018). On the elusive nature of critical (accounting) research. *Critical Perspectives on Accounting*, 50, 1–12.

Gottlieb, U., Johed, G., & Hansson, H. (2022). Accounting and accountability for farm animals: Conceptual limits and the possibilities of caring. *Critical Perspectives on Accounting*, 84, 102409. doi:10.1016/j.cpa.2021.102409.

Granà, F., Dimes, R., Busco, C., & de Villiers, C. (2025). Addressing systemic social and environmental challenges: the role of accounting and accountability practices. *Accounting, Auditing & Accountability Journal*, 38(5), 1325–1346.

Gunderson, R. (2013). From cattle to capital: Exchange value, animal commodification and barbarism. *Critical Sociology*, 39(2), 259–275. doi:10.1177/0896920511421031.

Hamilton, C. (2017). *Defiant Earth: The fate of humans in the Anthropocene*. Polity.

Haraway, D. J. (2008). *When species meet*. University of Minnesota Press.

Himick, D., & Vinnari, E. (2023). Counter accounts, accountability and governance. In G. Carnegie & C. Napier (Eds), *Handbook of accounting, accountability and governance* (pp. 349–368). Edward Elgar.

Malm, A. (2016). *Fossil capital: The rise of steam power and the roots of global warming*. Verso Books.

Maroun, W., Ecim, D., & Cerbone, D. (2023). Refining integrated thinking. *Sustainability Accounting, Management and Policy Journal*, 14(7), 1–25. doi:10.1108/SAMPJ-07-2021-0268.

Marx, K. (1867/1976). *Capital: A critique of political economy* (Vol. I). Penguin Classics.

McLaren, J., & Appleyard, T. (2022). Social movements, identity and disruption in organizational fields: Accounting for farm animal welfare. *Critical Perspectives on Accounting*, 84, 102310. doi:10.1016/j.cpa.2021.102310.

Miller, P., & Power, M. (2013). Accounting, organizing and economizing: Connecting accounting research and organization theory. *Academy of Management Annals*, 7 (1), 557–605. doi:10.5465/19416520.2013.783668.

Milne, M. J., & Gray, R. (2013). W(h)ither ecology? The triple bottom line, the global reporting initiative, and corporate sustainability reporting. *Journal of Business Ethics*, 118(1), 13–29.

Pesci, C., Gelmini, L., & Vola, P. (2024). On the accounting implications of the dilemma: Who speaks for nature? *Accounting, Auditing & Accountability Journal*, 37(9), 75–99. doi:10.1108/AAAJ-06-2022-5890.

Sahlins, M. (1972/2013). *Stone age economics*. Routledge.

Tavakolnia, E. (2023). Making animals visible in sustainability accounting with a critical look at financial valuation. *International Journal of Management, Accounting & Economics*, 10(5), 360–378. doi:10.5281/zenodo.8062966.

Vinnari, E., & Laine, M. (2017). The moral mechanism of counter accounts: The case of industrial animal production. *Accounting, Organizations and Society*, 57, 1–17. doi:10.1016/j.aos.2017.01.002.

Vinnari, E., & Vinnari, M. (2022). Making the invisibles visible: Including animals in sustainability (and) accounting. *Critical Perspectives on Accounting*, 82, 102324. doi:10.1016/j.cpa.2021.102324.

Vollmer, H. (2021). Public value and the planet: Accounting in ecological reconstitution. *Accounting, Auditing & Accountability Journal*, 34(7), 1527–1554. doi:10.1108/AAAJ-11-2019-4283.

Vosselman, E. (2022). The performativity of accounting: advancing a posthumanist understanding. *Qualitative Research in Accounting & Management*, 19(2), 137–161.

Wadiwel, D. (2015). *The war against animals*. Brill.

Wadiwel, D. (2020). The working day: Animals, capitalism and surplus time. In C. Blattner, K. Coulter & D. Kymlicka (Eds), *Animal labour: A new frontier of inter-species justice* (pp. 181–206). Oxford University Press. doi:10.1093/oso/9780198846192.001.0001.
Wadiwel, D. (2023). *Animals and capital*. Edinburgh University Press. doi:10.3366/edinburgh/9781399518062.001.0001.
Wolfe, C. (2003). *Animal rites: American culture, the discourse of species and post-humanist theory*. University of Chicago Press.
Yusoff, K. (2018). *A billion black Anthropocenes or none*. University of Minnesota Press.

11

ANIMALS AND POLICY AND PLANNING

Introduction

Policy and planning are central to responsible business because they shape how ethical commitments, sustainability goals and social responsibilities are translated into practice. Policies reflect societal priorities and can establish higher standards that define acceptable and unacceptable business conduct in relation to society, the environment and animals, among other issues of concern (see Coulter, 2016). These policies emerge from governance processes involving governments, non-governmental organisations, firms and other stakeholders, establishing the normative and regulatory framework within which businesses operate (see Gond et al., 2011; Moon & Vogel, 2009). Planning complements policy by focusing on how these commitments are implemented. In a responsible business context, planning involves integrating sustainability policies into corporate strategy, operations and decision-making, including risk and supply chain management. Policy and planning help move responsible business beyond voluntary intentions towards institutionalised practices, enabling firms to align their activities with broader societal goals and to respond more effectively to complex environmental and social challenges. Drawing on the work of David Fennell and insights from the Centre for Animal-based Tourism, this chapter argues for a more inclusive approach to policy and planning that integrates animals and their welfare into responsible business governance.

Although policy and planning discussions have increasingly addressed environmental concerns and biodiversity (Coffey et al., 2023; Schneider & Ingram, 1997), there remains a significant gap in frameworks that recognise animals as having intrinsic value and legitimate stakeholder status. For example, as highlighted in Chapter 1, the 17 Sustainable Development Goals (SDGs), which play a central role in shaping global sustainable development policy and

DOI: 10.4324/9781003585541-16

planning, fail to recognise animals as stakeholders or rights-bearing entities. While SDG 14 (Life Below Water) and SDG 15 (Life on Land) focus on protecting marine and terrestrial ecosystems, conserving biodiversity and preventing species extinction, animals are treated primarily as components of ecological systems rather than as sentient individuals whose welfare and rights warrant ethical consideration. Similarly, SDG 12 (Responsible Consumption and Production) and SDG 2 (Zero Hunger) address animals mainly in relation to sustainable agriculture, fisheries and food systems, thereby framing them as resources within largely anthropocentric development objectives (see e.g. Fennell & Sheppard, 2021; Kopnina, 2020; Tallberg & Tienari, 2022).

A comparable pattern can be observed in the Global Code of Ethics for Responsible Tourism (United Nations World Tourism Organisation (UNWTO), 1999), which serves as an important policy framework for the development of responsible tourism policies and standards. Most of the prescriptions contained in the Code's ten Articles and accompanying provisions emphasise human rights, freedoms and benefits, with comparatively limited attention to environmental concerns and even less to the ethical treatment of animals (UNWTO, 1999). The Code's prevailing anthropocentric orientation undermines its potential as a truly responsible ethical framework, as it fails to acknowledge the needs and interests of the millions of animals used within the tourism industry for human enjoyment and benefit (Fennell, 2014).

These two global policy frameworks underscore the need to expand the moral circle and adopt broader governance perspectives that recognise multiple stakeholders, including nonhuman actors. Such a shift challenges long-standing assumptions that policy and planning for responsible business are inherently human-centred and primarily driven by socioeconomic objectives. In areas such as biodiversity conservation and animal welfare, scholars increasingly argue that responsible business requires a rethinking of policy and planning frameworks to account for ecological limits, moral responsibility and principles of multispecies justice (e.g. Coulter, 2016; Sheppard & Fennell, 2019). From this perspective, this chapter calls for a more inclusive approach to policy and planning that aligns business activity with the wellbeing of both human and nonhuman stakeholders, thereby expanding the scope of responsible business beyond anthropocentric concerns. In doing so, this chapter contributes to the book's broader project of developing an interspecies approach to responsible management by showing how governance structures – not just firm-level practices – shape whose lives are protected and whose are marginalised.

Our expert collaborators and the case

This chapter is based on an interview with Professor David Fennell, who holds a PhD in Geography from the University of Western Ontario, as well as an MA in Recreation and Leisure Studies and a BES in Environment and Resource Studies from the University of Waterloo. He is a Fellow of the

International Academy for the Study of Tourism and holds adjunct academic appointments internationally. His research focuses on animal-based tourism, ecotourism, sustainable tourism and tourism ethics. A central thrust of his work involves applying philosophical perspectives and theories from other disciplines to address some of tourism's most persistent challenges. David has authored and edited numerous influential books, including *Ecotourism, Tourism Ethics, Sustainable Tourism* and *Tourism and Animal Ethics*. He is the founding Editor-in-Chief of the *Journal of Ecotourism*, editor of a Routledge book series on tourism ethics and editor of two Routledge handbooks: *Tourism and the Environment* and *Ecotourism*. Over the past 25 years, he has been the principal force shaping the field of tourism and animal ethics, a subfield he established in 2000. His interest in policy and planning emerged from a critical examination of the categorisation of billfish tourism as a form of ecotourism (Fennell, 2000). Through this work, he recognised that the absence of clear definitions and robust ecotourism policies hindered the responsible planning, development and management of this emerging tourism activity.

David also serves as the Executive Director of the Centre for Animal-based Tourism (CAT, www.catconsortium.com), which seeks to disrupt and transform prevailing knowledge and practice related to the use of animals in tourism. CAT does so by integrating expertise in animal welfare and ethics, ecotourism, education and literacy and conservation. Research across these four core components is intended to drive meaningful change in how animals are used within tourism contexts. Achieving this goal requires robust policy and planning frameworks that support the effective operationalisation of animal–human principles and practices for the benefit of both. Animals play a vital role in the global tourism industry, appearing in captive settings such as zoos, in the wild through activities like whale watching and as active participants in tourism experiences such as horseback riding and dog sledging (Fennell, 2012; Rickly & Kline, 2021). One of CAT's central objectives is to provide knowledge and leadership to national and regional policymakers in shaping the future of tourism planning, development and management. This chapter builds on David's work as a scholar who is not only engaged in the production of knowledge on more humane animal-tourism practices but also an active advocate for tourism policies that recognise animals as key tourism stakeholders. By using the case of CAT, we examine David's contributions to the field, illustrating how his research advances a more inclusive understanding of policy and planning that accounts for both human and nonhuman species. In doing so, the chapter calls for future research that moves beyond anthropocentrism, positioning policy and planning as instruments that acknowledge the interests and needs of nonhuman animals. For responsible managers, these governance structures shape the regulatory environments in which firms operate. Understanding them is therefore essential for embedding animal-inclusive practices within organisational strategy and decision-making.

Conceptual foundations: What are policy and planning?

Policy and planning are closely interconnected yet conceptually distinct dimensions through which governments organise society, regulate conflict, influence behaviour and distribute benefits and burdens (Dye, 2013). Policy refers to the formal rules, strategies and guiding principles adopted by societies to address public issues and steer collective action. As emphasised by Howlett et al. (2020), policy should not be understood as a single decision or static document, but rather as a complex and ongoing process through which governments, institutional actors and other stakeholders respond to economic, social and environmental challenges. Their influential policy cycle model – comprising agenda setting, policy formulation, implementation and evaluation – highlights how policy evolves through multiple stages and involves a wide range of actors, interests and instruments. This process-oriented understanding of policy has been widely adopted across disciplines such as environmental studies, tourism and responsible business, where attention to implementation and outcomes is essential.

Planning, by contrast, concerns the processes through which these policy intentions are translated into coordinated action over time and across space, including the allocation of resources, the design of interventions and the management of implementation (Faludi, 1973; Hall, 2002). This perspective emphasises that planning is not merely a technical or administrative activity, but an inherently normative and political practice, embedded in value judgments about priorities, interests and the distribution of resources. As a result, planning both reflects and reproduces broader social assumptions, including those related to human–nonhuman relations, making it a particularly important site for critical inquiry into ethical considerations and challenges to anthropocentric modes of governance. Together, policy and planning provide an analytical framework for understanding how social and environmental commitments are enacted, adapted and contested in practice.

Policy and planning in responsible business

While a voluminous body of corporate social responsibility (CSR) literature has framed responsibility in largely instrumental terms – as a firm-level strategy and voluntary initiative aimed at responding to primary stakeholders, while leaving policymaking and planning functions to democratically governed public institutions within a limited state (see Barnett, 2019; Mäkinen, 2026; McWilliams & Siegel, 2001) – a growing body of research has increasingly examined policy and planning as central components of responsible and ethical business conduct. This more expansive understanding of CSR conceptualises responsibility not merely as discretionary corporate action, but as companies' ethical role in addressing regulatory gaps by introducing deliberative procedures and making political decisions to manage their impacts on society and the environment (see Gond et al., 2011; Mäkinen, 2026). In line with this perspective, political institutions such as the European Union have explicitly emphasised the role and responsibility of firms in addressing their social and environmental impacts through regulatory and policy frameworks (European Commission, 2011). Consequently, the traditional understanding of CSR as a voluntary activity has been progressively replaced by an understanding of CSR as embedded within national and international institutional contexts, shaped by policy, planning and governance arrangements (Knudsen et al., 2015; Moon & Vogel, 2009).

Some scholars have highlighted the increasing involvement of business organisations in governance functions that were traditionally reserved for states (Sherer & Palazzo, 2011). For example, the literature on political CSR examines how firms engage in policymaking, standard-setting and planning processes, particularly in contexts where state regulatory capacity is limited (see Frynas & Stephens, 2015; Mäkinen, 2026). Research on sustainable supply chains and CSR implementation has likewise placed strong emphasis on planning, demonstrating that CSR commitments require both policies and planning across organisational and supply-chain boundaries (e.g. Gond et al., 2011; Sherer & Palazzo, 2011). In this body of work, planning is understood as the mechanism through which CSR policies, such as labour and environmental standards, are operationalised through sourcing strategies and monitoring systems. These studies show that, in the absence of deliberate and coordinated planning, CSR policies often remain symbolic or become decoupled from everyday business practice.

Across industries such as mining and tourism, there has been growing concern about the underrepresentation of marginalised stakeholders in policy and planning processes (e.g. Abdullah et al., 2022; Roussey et al., 2022). From a policy perspective, this includes actors who are prevented from, or unable to, participate meaningfully in decision-making related to planning and development. Among these marginalised stakeholders are nonhuman species, whose interests are rarely acknowledged or represented within formal policy frameworks (Abdullah et al., 2022; Coulter, 2016; Sheppard & Fennell, 2019). Although there have been positive developments across the world – such as the Nova Scotia government's approval of plans for Canada's first-ever whale sanctuary (Mitchell, 2025) and Spain's adoption of legislation recognising animals as sentient beings whose rights and interests must be taken into account (Hermida & Sánchez, 2021) – there remains a need for a deeper understanding of how animal interests are incorporated into policymaking and planning processes at the intersection of business and society. This issue is taken up in the next chapter.

Moving towards interspecies policy and planning

Drawing on the case of the CAT as an organisation dedicated to developing responsible and innovative solutions to improve the welfare of animals in tourism, as well as insights from the interview with David, the following section outlines three key priorities: literacy and knowledge translation; advancing animal welfare and other ethical domains; and applying global principles to local practices. Together, these priorities, based on David's research, are presented as essential for advancing policy and planning frameworks that support interspecies wellbeing. By elaborating on these priorities, the section highlights pathways for influencing those in positions of power to develop policies and formal regulations that reduce the marginalisation of animals in business and society.

Literacy and knowledge translation

While literacy is often defined in terms of possessing basic functional skills such as reading and writing (Keefe & Copeland, 2011), the Cambridge Dictionary (online) defines it as competence or knowledge in a particular domain. To foster a society and industries that demonstrate greater responsibility towards animals and other species, managers, policymakers and other stakeholders need to understand the extent to which their actions affect the wellbeing of animals and more-than-human life (see Fennell, 2022b). From this perspective, educating key industry stakeholders and sectors about animals and human relationships with other species is crucial for developing policies and planning processes that account for both human and nonhuman entities. While knowledge in a business context is often framed as a form of intellectual capital that enhances and sustains competitive advantage (Grant, 1996; Yiu & Law, 2014), such an instrumental view is insufficient to support greater responsibility in business and society. Knowledge here must also be understood as a normative and ethical resource, oriented towards advancing the wellbeing of humans, animals and the planet (see Fennell, 2022a).

When discussing knowledge, two forms of knowledge production can be distinguished: one generated within institutions of higher education and organised around specific disciplines and fields, and another produced by industry, government and consultants, which is oriented towards addressing practical problems and real-world issues (Gibbons et al., 1994; Tribe, 1997). Considering these two forms of knowledge, scholarship adopting an animal perspective on responsible business must engage meaningfully with practitioners and policymakers, so that academic knowledge can be translated into actionable insights that enable these actors to recognise animals as legitimate stakeholders (Fennell, 2014, 2022a; Sheppard & Fennell, 2019). From this perspective, knowledge translation is understood as a dynamic, iterative process that involves the synthesis, dissemination, exchange and ethically grounded application of knowledge to improve policy, planning, development and managerial practices (see Fennell, 2022a; Straus et al., 2009).

To disrupt and transform prevailing knowledge and questionable practices surrounding the use of animals in tourism, CAT puts into practice the Tourism Knowledge Translation (TKT) framework developed by Fennell (2022a). Through this approach, CAT seeks to improve the living and working conditions of animals by increasing animal welfare literacy among all tourism stakeholders and by translating academic research into accessible and practical guidance for practitioners. At its core, the TKT framework understands knowledge as something that must be adapted, communicated and co-produced across diverse stakeholder groups. Central to this approach is an emphasis on engagement and relationships, which is particularly important in animal-based sectors such as tourism, where ethical complexity and contested values require ongoing dialogue rather than prescriptive solutions (see Fennell, 2022a). These ideas are reflected in CAT's commitment to creating responsible, cutting-edge solutions that bridge the gap between scholarly knowledge and industry application, ensuring that new insights reach those who shape tourism practice as quickly as possible. Additionally, CAT's emphasis on collaboration with industry, NGOs, governments and academic affiliates underscores its role in supporting research-based leadership to national and regional policymakers, thereby contributing to more informed animal-based tourism planning, development and management that better accounts for animal wellbeing and ethical considerations in decision-making.

Animal welfare and beyond

Approaches to policy and planning involving animals must be sensitive to the economic sector in question and to the cultural and institutional contexts within which change is needed. In sectors where animals are deeply embedded in economic activity, policies are more likely to gain traction when they begin with an animal welfare perspective rather than an immediate shift towards animal rights. While animal welfare and animal rights are often used interchangeably, they represent fundamentally different ethical positions. The distinction is commonly illustrated through the cage analogy: animal rights call for empty cages, whereas animal welfare advocates for larger or better cages (Sheppard & Fennell, 2019, p. 135). Within tourism ethics research, most scholarship has focused on animal welfare rather than animal rights, implicitly accepting the use of animals for tourism purposes provided their wellbeing is safeguarded (see Fennell, 2013; Rickly & Kline, 2021). From this perspective, ethical responsibility is understood as ensuring that animals are adequately fed and watered, appropriately housed, free from pain and suffering and able to engage in species-appropriate behaviours (Bekoff & Nystrom, 2004).

In the interview, David explains that adopting an animal rights position as a starting point for policy and planning would entail the immediate elimination of all forms of animal-based tourism, a position unlikely to be politically feasible or socially acceptable within current governance systems. Indeed, according to animal rights, animals have moral and legal rights that include the right not to be harmed and that animals' lives are valuable in and of themselves (intrinsic value) – their lives are not valuable because of what they can do for humans (instrumental value) (Bekoff & Nystrom, 2004). Such an ethical perspective would limit engagement with policymakers and practitioners, risking the foreclosing of incremental improvements that can meaningfully enhance animal wellbeing. By contrast, an animal welfare approach is more acceptable in industries such as tourism, enabling the development of policies, standards and plans that improve animal conditions while keeping channels open for dialogue, education and institutional change. This welfare-first strategy is evident in CAT's work, which seeks to raise animal welfare literacy, influence policy and progressively shift norms and practices within the tourism industry.

According to David, the progression from animal welfare towards animal rights mirrors debates in environmental ethics and sustainability. Arne Naess's (2005) distinction between shallow environmentalism (anthropocentric, human-focused resource management) and deep ecology (ecocentric), highlights that transformative ethical shifts often require intermediate stages. Sustainable development has frequently been understood as a pragmatic pathway that moves societies away from narrow anthropocentrism towards deeper ecological awareness, even if it does not fully embody deep ecology itself. Similarly, in relation to animals, animal welfare can function as a necessary transitional ethical framework that challenges exploitative practices, embeds moral consideration into policy and planning and gradually shifts the status of animals from moral objects to beings with direct moral standing and basic rights. From this perspective, animal welfare should not be seen as an endpoint, but as a critical stepping stone towards more robust ethical frameworks that may eventually encompass other ethical domains (e.g. justice, virtue ethics, care ethics) if that means greater protection of animals. Policy and planning processes that begin with welfare-oriented reforms create the institutional and cultural conditions for ethical pluralism under which more fundamental shifts in how animals are valued and included in business and society can occur. In this way, advancing animal welfare is not a compromise of ethical ambition, but a strategic and normative pathway towards more just and inclusive multispecies futures.

Applying global principles to local practices

One aspect highlighted by David in the interview was that when developing policy and planning aiming to support the wellbeing of different species in business and society, we need to consider both the global and local contexts. Indeed, despite ongoing efforts to improve practices in animal-based tourism, structural challenges continue to limit meaningful progress towards enhancing animal welfare around different regions of the world. Many of these challenges are deeply embedded in the cultural and political tensions that arise between globally promoted ethical principles and the values, traditions and economic realities of local operators (Fennell, 2022c). As Duffy and Moore (2011) observe, global governance actors often seek to shift decision-making power away from host nations towards international institutions, a dynamic that can generate resistance and undermine locally grounded implementation. Similarly, von Essen et al. (2020) highlight the dual challenge posed by cultural relativism and the uncritical imposition of universal standards, both of which can obstruct effective and legitimate animal welfare reforms. Addressing these tensions requires a multifaceted approach that includes strengthened animal welfare legislation, the development of best-practice guidelines, certification schemes and codes of conduct, alongside greater efforts to educate tourists and deepen understanding of cultural differences (von Essen et al., 2020). The gap between global ethical frameworks and local practices ultimately contributes to the failure of existing governance arrangements to adequately protect the interests and welfare of animals used in tourism (Fennell, 2022c).

One example of CAT's innovative research agenda is a planned project that seeks to address the complex and prevailing ethical challenges associated with the use of elephants in tourism in Asia. To be effective, the project recognises the need to engage with the multiple, interrelated dimensions that shape elephant-based tourism, including political and governance structures, economic dependencies and alternative livelihood strategies for local communities, protected area management, sociocultural traditions and the practices of animal-based tourism venues. Rather than isolating animal welfare as a technical issue, the project adopts a holistic perspective that situates elephant use within broader social, economic and institutional contexts. Central to this approach is the application of integrative social contracts theory to reconcile global ethical principles with locally embedded practices (see Carroll, 2004; Donaldson & Dunfee, 1994; Fennell, 2018, 2022c). By acknowledging the legitimacy of local norms while upholding overarching moral standards, this framework provides a pathway for developing policy and planning solutions that support both improved elephant welfare and conservation outcomes, while remaining sensitive to local livelihoods and cultural values. To that end, the project follows the hypernorms for animal-based tourism (Fennell, 2022c), which hold that global principles should take precedence over local practices when the latter conflict with fundamental ethical obligations. These hypernorms – which include the pinnacle of practice, the principles of beneficence and non-maleficence; enhanced literacy and education; justice; recognition of animal agency and self-determination; and animal-informed consent – play a critical role in shaping practical governance mechanisms that align elephant conservation and welfare with more sustainable and socially just forms of tourism development.

We concluded that advancing interspecies policy and planning requires a gradual, context-sensitive approach that balances ethical ambition with practical feasibility. Beginning with an animal welfare orientation provides a politically and socially viable entry point for improving animal wellbeing in industries where animal perspectives are absent or only beginning to emerge, enabling the development of policy and planning frameworks within existing governance structures while fostering dialogue, learning and institutional adaptation. However, the welfare perspective should be understood as a transitional stage rather than an endpoint. Its purpose is to prepare the ground for broader ethical transformations that progressively expand the moral community towards animal rights and multispecies justice. Achieving such a progression demands careful attention to how universal ethical principles intersect with local cultural, economic and political realities – a dynamic illustrated in CAT's research initiatives, such as work on elephant-based tourism. By integrating ethical theory with context-aware policy and planning, these efforts demonstrate how responsible business can become more inclusive of different species. In doing so, they not only enhance animal welfare in the short term but also contribute to longer-term shifts in how animals are valued and treated within both business practices and society as a whole (Fennell et al., 2026).

Directions for future research

Future research could address the prevailing gap between globally articulated ethical principles and locally embedded cultural, economic and institutional practices. This disconnect frequently undermines efforts to protect animal interests and welfare (see Fennell, 2022c). This tension is particularly evident in sectors where animal use is culturally normalised and economically significant, making the direct imposition of global standards both politically contentious and practically ineffective. Building on Fennell's (2022c) application of integrative social contracts theory, future studies could further explore how localised micro-social contracts can be ethically justified within the boundaries of a global macro-social contract, allowing locally specific practices to continue, provided they do not violate overarching ethical obligations towards animals.

Extending this line of inquiry across different animal-dependent industries and diverse geographical and cultural contexts would generate valuable insights into the political, economic, social, legal and environmental conditions that shape the feasibility of interspecies policy and planning. Such research could help clarify the degree of flexibility that can be given to local actors without compromising the wellbeing of animals or the livelihoods of people involved in these industries. In addition, incorporating complementary ethical perspectives such as virtue ethics, ethics of care and ecofeminism would enrich current frameworks by foregrounding relationality, empathy and moral character, thereby expanding the normative foundations of policy and planning for interspecies wellbeing (see Fennell, 2024). Together, these approaches offer promising pathways for developing more context-sensitive, ethically grounded and practically effective governance mechanisms that move beyond abstract principles towards meaningful protection for both human and nonhuman stakeholders.

As Fennell (2022b, p. 9) also notes, future research could usefully compare levels of animal welfare literacy among organisations working with the same species (e.g. different elephant-tourism operators) as well as across species (e.g. cetaceans versus great apes). Such comparative analyses would help illuminate sector- and species-specific gaps in animal welfare knowledge and practice, along with their implications for interspecies policy and planning. This also raises the broader question of how a business model reliant on the use of animals can be transformed into one that operates entirely without them. The case of Cirque du Soleil provides an interesting example of how circus business models, traditionally dependent on animal performances, can be challenged and reconfigured into successful, animal-free enterprises. Future studies could further explore how policy and planning might be developed and implemented to support transitions from an animal welfare orientation towards approaches that more explicitly stress an animal rights perspective.

Ethical implications

A shift towards greater interspecies inclusivity, as informed by the priorities discussed in this chapter, carries significant ethical implications for how animals are conceptualised and positioned within the field of responsible business. Framing animal welfare as an initial basis for policy and planning reflects a pragmatic ethical stance that prioritises reducing the suffering animals may experience across economic sectors. This approach recognises that animals warrant moral consideration and affirms the responsibility to safeguard their fundamental needs when they are involved in human activities (Fennell, 2022b; Fennell & Sheppard, 2021). Ethically, it positions welfare-oriented policies as incremental expressions of moral responsibility while simultaneously expanding the moral circle, challenging anthropocentric assumptions by acknowledging animals as stakeholders whose wellbeing merits protection through formal governance mechanisms (Sheppard & Fennell, 2019).

The openness to consider animal welfare alongside other ethical theoretical perspectives raises critical questions about moral obligation, justice and the legitimacy of certain business practices (see Fennell & Sheppard, 2021). Treating welfare as a transitional framework implies an ethical commitment to ongoing moral reflection and transformation rather than maintaining the status quo. This trajectory aligns with broader debates in animal ethics, environmental ethics and sustainability, where pragmatic reforms are often understood as pathways towards more far-reaching ethical reorientations. Consequently, it requires policymakers, managers and scholars to remain attentive to power relations, cultural contexts and the voices – human and nonhuman – that are marginalised in decision-making and affected by the restructuring of established business systems. Although the priorities discussed in this chapter aim to minimise harm to animals within various organisational contexts, they nevertheless accept the use of animals within existing business and societal systems – a practice that remains ethically problematic from an animal rights perspective. This acceptance can be seen as a necessary, albeit ethically contentious, step towards developing more robust policy and planning frameworks that ultimately support more substantive transformations in human–animal relations.

Practical implications for responsible management

This chapter offers practical implications for both policymakers and business organisations committed to advancing more responsible and ethical forms of business through interspecies policy and planning. It highlights the marginalisation of animals and other nonhuman species within dominant global policy frameworks and demonstrates how this omission can have direct effects on national and regional policies. For example, the highly anthropocentric orientation of the Global Code of Ethics for Tourism (UNWTO, 1999) continues to shape the priorities of tourism strategies and regulatory frameworks worldwide, often privileging human rights, economic growth and environmental concerns while giving limited attention to animal welfare. This influence is evident in contexts like Canada, where tourism policy strongly emphasises sustainability and nature-based tourism yet affords minimal consideration to the welfare of animals and other nonhuman species involved in tourism activities (Sheppard & Fennell, 2019).

For policymakers and business organisations seeking to develop more ethical governance frameworks, a key lesson emerging from this chapter is the importance of recognising animals as legitimate stakeholders and animal welfare as a core responsibility of responsible business – particularly in industries that rely heavily on the use of animals. The development and effectiveness of interspecies policy and planning depend not only on strong ethical commitments from diverse stakeholders but also on high levels of animal welfare literacy, sensitivity to cultural and institutional contexts and a nuanced understanding of how global ethical principles are interpreted and enacted within local practices (see Fennell, 2022b, 2022c). The chapter also underscores the value of sustained dialogue and collaboration between practitioners, policymakers and scholars, as research-informed engagement can support the translation of ethical knowledge into actionable policies and planning processes (see Fennell, 2022a).

Finally, this chapter cautions against treating animal welfare as an ethical endpoint. Instead, it positions welfare-oriented reforms as part of a broader ethical trajectory that challenges entrenched anthropocentric assumptions and opens space for innovation in governance, planning and business practice. From this perspective, interspecies policy and planning is best understood as an ongoing process of ethical deliberation, learning and reflexivity – one that progressively reshapes industries and how humans coexist with other species in more just, transparent and sustainable ways. Policy and planning, therefore, are not neutral administrative tools but ethical instruments that determine whether animals are recognised as stakeholders in sustainable and responsible governance.

Reflective activities inside/outside the classroom

A Revision questions that test knowledge ("what?")

- A1 According to the chapter, how has policy and planning been conceptualised in relation to responsible business and what key limitations have been identified in this context?
- A2 What are the key elements of policy and planning?
- A3 How does the case of CAT and a tourism policy context contribute to our understanding of a less anthropocentric view of policy and planning?
- A4 What is meant by interspecies policy and planning and what new insights does this approach bring to discussions of policy and planning within a responsible business context?

B Reflective questions that build understanding ("so what?")

- B1 Why is it important to consider animals and other nonhuman species in policy and planning?
- B2 How can policy and planning be leveraged to support the wellbeing of both humans and nonhuman species?
- B3 Why is it essential to move beyond an anthropocentric approach to policy and planning in discussions of responsible business?

- B4 Why should nonhuman species be recognised as key stakeholders in policy and planning processes?

C Take action ("do")

- C1 What factors should be considered when developing policy and planning that are inclusive of multiple species?
- C2 How can policy and planning contribute to increasing animal welfare literacy and awareness of animal rights?
- C3 What policies could be developed in my university or workplace to support the wellbeing of nonhuman species?
- C4 How might I contribute to the design and implementation of policy and planning that benefits nonhuman species?

D Ideas to discuss with colleagues, friends and family ("share")

- D1 To what extent is it morally acceptable to allow the use of animals, provided appropriate measures are taken to ensure their basic needs are met?
- D2 Can an animal-welfare ethical perspective be considered sufficient for informing interspecies policy and planning? Why or why not?
- D3 Are there examples within your field where an organisation has adopted policies or planning that align with the interspecies perspective discussed in this chapter?
- D4 Have you ever reflected on the limited attention given to animals in policy and planning? What are your thoughts on the marginalisation of animals?

Key ideas

Animal welfare

Animal welfare is defined as the state of animals in relation to their ability to cope with their environment, permitting their use provided that appropriate measures are taken to ensure their basic needs are met. Animal welfare should guarantee five freedoms: freedom from hunger and thirst; freedom from discomfort; freedom from pain, injury and disease; freedom to express normal behaviour; and freedom from fear and distress (see Fennell & Sheppard, 2021; Hill & Broom, 2009).

Animal rights

Animal rights holds that individual animals possess inherent value and should be regarded as ends in themselves, rather than as means to human ends, in much the same way that humans are understood to have intrinsic moral worth. One element of this is that animals meet key moral criteria, including consciousness, intentionality in action and thought and sentience – that is, the capacity to experience states such as pain and pleasure (Regan, 2004; Sheppard & Fennell, 2019).

Integrative social contract theory

Integrative theory recognises ethical obligations based upon two levels of consent; first, to a theoretical "macrosocial" contract appealing to all rational contractors, and second, to real "microsocial" contracts by members of numerous localised communities (Donaldson & Dunfee, 1994).

Hypernorms

Hypernorms refer to principles so fundamental that, by definition, they serve to evaluate lower-order norms, reaching to the root of what is ethical for humanity. They represent norms by which all others are to be judged (Donaldson & Dunfee, 1994; Fennell, 2022c).

Interspecies governance

Interspecies governance refers to policy and planning frameworks that recognise animals as legitimate stakeholders whose interests warrant formal consideration within regulatory and institutional decision-making. It challenges anthropocentric governance by expanding the moral community to include nonhuman beings (Coulter, 2016; Sheppard & Fennell, 2019).

References

Abdullah, T., Lee, C., & Carr, N. (2022). Conceptualising human and non-human marginalisation in tourism. *Tourism and Hospitality Research*, 23(2), 254–268. doi:10.1177/14673584221089735.

Barnett, M. L. (2019). The business case for corporate social responsibility: A critique and an indirect path forward. *Business & Society*, 58(1), 167–190. doi:10.1177/0007650316660044.

Bekoff, M., & Nystrom, J. (2004). The other side of silence: Rachel Carson's views of animals. *Zygon*, 39(4), 861–883. doi:10.1111/j.1467-9744.2004.00625.x.

Coffey, B., Damiens, F. L. P., Hysing, E., & Torabi, N. (2023). Assessing biodiversity policy designs in Australia, France and Sweden. Comparative lessons for transformative governance of biodiversity? *Journal of Environmental Policy & Planning*, 25 (3), 287–300. doi:10.1080/1523908X.2022.2117145.

Coulter, K. (2016). *Animals, work and the promise of interspecies solidarity*. Palgrave Macmillan. doi:10.1057/9781137558800.

Donaldson, T., & Dunfee, T. W. (1994). Toward a unified conception of business ethics: Integrative social contracts theory. *The Academy of Management Review*, 19(2), 252–284. doi:10.2307/258705.

Duffy, R., & Moore, L. (2011). Global regulations and local practices: The politics and governance of animal welfare in elephant tourism. *Journal of Sustainable Tourism*, 19(4–5), 589–604. doi:10.1080/09669582.2011.566927.

Dye, T. R. (2013). *Understanding public policy* (14th ed.). Pearson Education.

European Commission. (2011, October 25). *A renewed EU strategy 2011–14 for corporate social responsibility*. COM (2011) 681 final, Brussels.

Faludi, A. (1973). *A reader in planning theory*. Pergamon Press. doi:10.1016/B978-0-08-017066-4.50005-9.

Fennell, D. A. (2000). Comment: Ecotourism on trial: The case of billfish angling as ecotourism. *Journal of Sustainable Tourism*, 8(4), 341–345. doi:10.1080/09669580008667370.

Fennell, D. A. (2012). *Tourism and animal ethics*. Routledge. doi:10.4324/9780203153642.

Fennell, D. A. (2013). Tourism and animal welfare. *Tourism Recreation Research*, 38 (3), 325–340. doi:10.1080/02508281.2013.11081757.

Fennell, D. A. (2014). Exploring the boundaries of a new moral order for tourism's global code of ethics: An opinion piece on the position of animals in the tourism industry. *Journal of Sustainable Tourism*, 22(7), 983–996. doi:10.1080/09669582.2014.918137.

Fennell, D. A. (2018). The future of ethics in tourism. In C. Cooper & E. Fayos Sola (Eds.), *The future of tourism* (pp. 155–178). Springer. doi:10.1007/978-3-319-89941-1_8.

Fennell, D. A. (2022a). The tourism knowledge translation framework: Bridging the canyon between theory and practice. *Current Issues in Tourism*, 25(5), 674–691. doi:10.1080/13683500.2021.1887099.

Fennell, D. A. (2022b). An animal welfare literacy framework for tourism. *Annals of Tourism Research*, 96, 103461. doi:10.1016/j.annals.2022.103461.

Fennell, D. A. (2022c). Bridging the global–local animal-based tourism divide. *Annals of Tourism Research*, 96, 103459. doi:10.1016/j.annals.2022.103459.

Fennell, D. A. (2024). Empathy in animal-based tourism contrasting constructed care and care ethics at a captive wildlife venue. *Tourism Recreation Research*, 49(6), 1222–1237. doi:10.1080/02508281.2023.2226038.

Fennell, D. A., Rizzolo, J. B., Theerasak, W., Pookkaman, N., & Ittiphong, K. (2026). Tourism at the intersection of animal ethics, human ecology and conservation: Asian elephants in violation? *Frontiers in Ethology*, 4, 1704639. doi:10.3389/fetho.2025.1704639.

Fennell, D. A., & Sheppard, V. (2021). Tourism, animals and the scales of justice. *Journal of Sustainable Tourism*, 29(2–3), 314–335. doi:10.1080/09669582.2020.1768263.

Frynas, G., & Stephens, S. (2015). Political corporate social responsibility: Reviewing theories and setting new agendas. *International Journal of Management Reviews*, 17(4), 483–509. doi:10.1111/ijmr.12049.

Gibbons, M. C., Limoges, H., Nowotny, S., Schwartzman, S., Scott, P., & Trow, M. (1994). *The new production of knowledge*. Sage.

Gond, J.-P., Kang, N., & Moon, J. (2011). The government of self-regulation: On the comparative dynamics of corporate social responsibility. *Economy and Society*, 14 (4), 640–671. doi:10.1080/03085147.2011.607364.

Grant, R. M. (1996). Toward a knowledge-based theory of the firm. *Strategic Management Journal*, 17(S2), 109–122. doi:10.1002/smj.4250171110.

Hall, P. (2002). *Cities of tomorrow: An intellectual history of urban planning and design in the twentieth century* (3rd ed.). Wiley.
Hermida, X., & Sánchez, E. (2021). Spain approves new law recognising animals as "sentient beings." *El País*. https://english.elpais.com/society/2021-12-03/spain-approves-new-law-recognizing-animals-as-sentient-beings.html.
Hill, S. P., & Broom, D. M. (2009). Measuring zoo animal welfare: theory and practice. *Zoo Biology*, 28(6), 531–544. doi:10.1002/zoo.20276.
Howlett, M., Ramesh, M., & Perl, A. (2020). *Studying public policy: Principles and processes* (4th ed.). Oxford University Press.
Keefe, E. B., & Copeland, S. R. (2011). What is literacy? The power of a definition. *Research and Practice for Persons with Severe Disabilities*, 36(3–4), 92–99. doi:10.2511/027494811800824507.
Knudsen, J. S., Moon, J., & Slager, R. (2015). Government policies for corporate social responsibility in Europe: A comparative analysis of institutionalization. *Policy & Politics*, 43(1), 81–99. doi:10.1332/030557312X656016.
Kopnina, H. (2020). Of elephants and men. *Animal Sentience*, 5(28), 2. doi:10.51291/2377-7478.1553.
Mäkinen, J. V. (2026). Political corporate social responsibility (PCSR). In C. Luetge & M. Thejls Ziegler (Eds), *Handbook of the philosophical foundations of business ethics*. Springer. doi:10.1007/978-94-007-6176-6_101-1.
McWilliams, A., & Siegel, D. (2001). Corporate social responsibility: A theory of the firm perspective. *Academy of Management Review*, 26, 117–127. doi:10.2307/259398.
Mitchell, B. (2025). Plans approved for Canada's first-ever whale sanctuary in Nova Scotia. *Blooloop*. https://blooloop.com/whale-sanctuary-canada-approved.
Moon, J., & Vogel, D. (2009). Corporate social responsibility, government and civil society. In A. Crane, A. McWilliams, D. Matten, J. Moon & D. S. Siegel (Eds), *The Oxford handbook of corporate social responsibility* (pp. 303–324). Oxford University Press. doi:10.1093/oxfordhb/9780199211593.003.0013.
Naess, A. (2005). The shallow and the deep, long-range ecology movement: A summary. In A. Drengson & H. Glasser (Eds), *Selected works of Arne Naess* (pp. 7–12). Springer.
Regan, T. (2004). *The case for animal rights*. University of California Press.
Rickly, J. M., & Kline, C. (2021). *Exploring non-human work in tourism: From beasts of burden to animal ambassadors*. De Gruyter. doi:10.1515/9783110664058.
Roussey, C., Balas, N., & Palpacuer, F. (2022). Political CSR initiatives as levers of marginalisation: The disconnect between representatives and the so-called represented in the mining industry. *Critical Perspectives on International Business*, 18(1), 31–49. doi:10.1108/cpoib-01-2018-0015.
Schneider, A., & Ingram, H. (1997). *Policy design for democracy*. University Press of Kansas.
Sheppard, V. A., & Fennell, D. A. (2019). Progress in tourism public sector policy: Toward an ethic for non-human animals. *Tourism Management*, 73, 134–142. doi:10.1016/j.tourman.2018.11.017.
Sherer, A. G., & Palazzo, G. (2011). The new political role of business in a globalized world: A review of a new perspective on CSR and its implications for the firm, governance and democracy. *Journal of Management Studies*, 48(4), 899–931. doi:10.1111/j.1467-6486.2010.00950.x.
Straus, A. E., Tetroe, J., & Graham, I. D. (2009). Defining knowledge translation. *Canadian Medical Association Journal*, 181(3–4), 165–168. doi:10.1503/cmaj.081229.
Tallberg, L., & Tienari, J. (2022). Human and nonhuman animals in a posthuman reality. In M. Sandberg & J. Tienari (Eds), *Transformative action for sustainable*

outcomes: Responsible organising (pp. 127–133). Routledge. doi:10.4324/9781003229728-23.

Tribe, J. (1997). The indiscipline of tourism. *Annals of Tourism Research*, 24(3), 638–657. doi:10.1016/S0160-7383(97)00020–0.

United Nations World Tourism Organisation (UNWTO). (1999). *Global code of ethics for responsible tourism*. www.untourism.int/global-code-of-ethics-for-tourism.

von Essen, E., Lindsjö, J., & Berg, C. (2020). Instagranimal: Animal welfare and animal ethics challenges of animal-based tourism. *Animals*, 10(10), 1830. doi:10.3390/ani10101830.

Yiu, M., & Law, R. (2014). Review and application of knowledge management and knowledge sharing in tourism. *Asia Pacific Journal of Tourism Research*, 19(7), 737–759. doi:10.1080/10941665.2013.812128.

12

ANIMALS AND COMPANY ACTIVISM

Introduction

Advocacy and lobbying for the protection of nonhuman animals can be traced back more than 150 years (Beers, 2006). The founding of the Royal Society for the Prevention of Cruelty to Animals (RSPCA) in 1824 was driven by public debates concerning the mistreatment of working animals, the ethics of scientific experimentation and blood sports such as fox hunting, bullbaiting and cockfighting. While the RSPCA is now recognised as an animal welfare organisation, it serves as historical evidence of early forms of activism aimed at improving the lives of animals within society. Early traces of animal advocacy can also be found in literature, such as in *Platero and I*, a novel by the Spanish writer Juan Ramón Jiménez, published in 1914. The story, which centres on the relationship between a child and a donkey, evokes empathy for animals while subtly criticising the mistreatment of working animals in society. Through anti-bullfighting statements, expressions of anguish over the castration of foals and forced pregnancies for milk production and reflections on the abandonment and abuse of domestic animals, the narrative conveys a deep concern for animal welfare, highlighting the emotional and ethical dimensions of human–animal relationships (Rodriguez García, 2024).

Both examples highlight the two dominant ethical perspectives in the public debate over animal activism: animal welfarism and animal rights. Welfarists accept the use of animals by humans provided it is free from cruelty and mistreatment, whereas animal rights advocates oppose any use of animals by humans, arguing that such practices violate animals' inherent rights and constitute a serious moral injustice (Beers, 2006; Schneeberger & Karimova, 2023). Activism grounded in animal rights ethics is based on the

DOI: 10.4324/9781003585541-17

principle of intrinsic value, which holds that animals possess inherent worth and should be treated as ends in themselves, not merely as means to human ends (see Fennell, 2014). While animal rights activism has often taken the form of anti-corporate movements challenging business models based on animal exploitation, there are also notable examples of companies advocating for animal rights principles. Lush, for example, is widely recognised for its strong commitment to animal advocacy, particularly through its opposition to animal testing (Aronczyk, 2013). The company enforces a strict policy prohibiting animal testing for both its products and ingredients and actively engages in public campaigns against animal cruelty. Similarly, The Body Shop became well known for its strong commitment to animal advocacy, particularly through its activism to ban animal testing in the cosmetics industry (Hartman & Beck-Dudley, 1999).

Some other companies have engaged in animal activism by adopting a welfarist perspective, advocating for the ethical treatment of animals while accepting their use in production systems (see García-Rosell, 2023). For example, while The North Face and Patagonia have acknowledged the use of animal-derived materials in their products, they have advocated for more humane practices and the ethical treatment of animals. Both companies have been active in greening their down supply chains, implementing standards and initiatives to improve animal welfare and drive positive change within the industry. Although animal advocacy and activism are reflected in the values and strategies of corporations, as illustrated previously, the literature on activism within business organisations has largely overlooked this dimension. Research has examined how activist campaigns influence companies to address social issues by strategically targeting firms and employing a diversity of tactics to gain public attention and pressure organisational decision-makers (Briscoe & Gupta, 2016). At the same time, scholars have begun to examine activism emerging from within businesses themselves, where CEOs or entire organisations take moral or political stances on particular issues, using their influence to pressure institutions and drive societal change. While activism within organisations has become a prominent topic in academic literature, the discussion remains largely anthropocentric in nature (García-Rosell, 2023). Drawing on the experience of Doris Schneeberger as a scholar and animal rights activist, this chapter argues for an interspecies approach to company activism grounded in anti-speciesism, transformation and interventionism.

Such an approach begins by recognising the limitations of existing understandings of company activism. The dominant focus on human-centred social and environmental causes has produced a narrow understanding of activism – one that largely ignores the suffering of animals directly or indirectly exploited by organisations. Industries such as agriculture, tourism, fashion and pharma rely heavily on animal labour or resources, yet theoretical discussions on the CEO (Chief Executive Officer) and corporate activism rarely address advocacy or political engagement on behalf of animals. This omission reinforces human exceptionalism in business education, implicitly teaching future managers that political action is warranted only when human rights or environmental issues are at stake (see Tallberg et al., 2022). Such a perspective not only overlooks animal exploitation but also normalises it, perpetuating the idea that animals fall outside the scope of ethical and political concern. By foregrounding this omission, the chapter reframes corporate activism as a site where responsibility is negotiated, reinforcing the book's call for a more inclusive approach to responsible management.

Our expert collaborators and the case

This chapter is based on an interview with Doris Schneeberger, who has a background in philosophy and holds two PhDs, one in animal ethics and one in management. Her research explores the role of nonhuman animals in organisations, drawing from organisational studies, animal ethics, psychoanalysis and the ethics of artificial intelligence. She is the author of *Envisioning a Better Future for Nonhuman Animals: Towards Future Animal Rights Declarations*, a book that presents ethically grounded arguments for animal rights (Schneeberger, 2024). The work combines internationally relevant perspectives with insights into the German-speaking academic and legal context and reflects on the potential for future universal declarations of animal rights. In her book, Doris discusses the interpretation and justification of animal rights, arguing that nonhuman animals are individuals with intrinsic value and that their welfare and interests deserve protection.

Doris is both an academic and a committed animal rights activist. Her passion for animals began early, leading her to adopt a vegetarian lifestyle in her teens and advocate for plant-based diets. She views her research as interventionist, aiming to introduce anti-speciesist perspectives to drive societal change. Part of this work is carried out through "Transfarmation Austria," a non-governmental organisation (NGO) she founded to promote a shift towards plant-based agriculture through education and support. The organisation helps farmers transition from livestock to sustainable alternatives like crop cultivation, educational farming and agrotourism. It also engages the public through workshops and events, highlighting the environmental, ethical and economic benefits of plant-based systems. Collaborating with researchers and policymakers, Transfarmation Austria supports evidence-based strategies for a resilient, ethical and compassionate food future.

This chapter builds on the work of Doris, both as a scholar and activist within Transfarmation Austria, using her contributions and the notion of transformation as a case study of how NGOs can promote a form of corporate activism among small agribusiness enterprises. Her work is contextualised alongside research from other scholars in the field to deepen the analysis. The chapter begins by defining the concepts of activism and corporate activism, followed by a critical examination of the anthropocentric limitations inherent in traditional approaches to corporate activism. It then presents a rationale for including animals in academic discussions of corporate activism, highlighting the practical, ethical and conceptual benefits of adopting a more inclusive framework. Drawing on the example of Transfarmation Austria, we explore Doris's work in detail, illustrating how this initiative expands and enriches our understanding of corporate activism by positioning animals not merely as passive subjects but as central to the activist agenda. Finally, the chapter advocates for future research and pedagogical practices that embrace a non-anthropocentric perspective, encouraging a rethinking of corporate activism that accounts for the interests and agency of nonhuman animals.

Conceptual foundations: What is activism?

Activism can be understood as a form of political action driven by moral convictions, in which the activist challenges the status quo on ethical grounds and uses that criticism to justify political intervention (Branicki et al., 2021). As such, it has traditionally been associated with grassroots movements, individuals and communities engaging in protests, demonstrations and campaigns to advocate for causes like human rights, environmental protection and social justice. At its core, activism is concerned with actions that seek to change the status quo regarding a social, political or environmental issue. It is a set of direct political actions informed by a moral position that criticises the current status quo and justifies intervention to transform existing structures within social, institutional and political systems (Saunders, 2022). Activism is not just about sporadic political action, but rather, it is embedded within a dense network of civic engagement (Youngs, 2019), expressed through a wide range of practices – from conventional activities such as debates, petitions and peaceful demonstrations to more confrontational tactics including civil disobedience, consumer boycotts and, in some cases, sabotage of private property. The nature of activism is diverse, not only in the methods employed but also in its intended targets and the issues under deliberation (Żuchowska-Skiba, 2026). These targets can include local citizens, municipal or national governments, international financial institutions and even broader systemic entities such as neoliberalism or globalisation.

Individuals who engage in activism are commonly referred to as "activists." Curtin and McGarty (2016) defined an activist as an individual who actively engages in efforts to promote social or political causes, often mobilising others in support of these causes. Activists can align with either the political left or right and may operate within large grassroots movements or within small NGOs (Curtin & McGarty, 2016). The term activist is not limited to individuals who actively participate in demonstrations, but also encompasses those who disseminate information, mobilise supporters and sustain the organisational structures that underpin social movements. Activists consist of a heterogeneous group defined by multiple dimensions. Activists may include students, social workers, academics and other professionals who engage in activism part-time, while others work full-time for organisations dedicated to advancing specific causes (Curtin & McGarty, 2016). As a result, activism is not confined to the margins of society as it permeates everyday life, institutions and even business organisations, challenging the boundaries of what is traditionally considered political action.

Activism in an organisational context

A growing body of literature examines activism within organisational contexts, investigating how companies and their leaders influence institutional and social change through political engagement (e.g. Corvellec & Stål, 2019; de Bakker, 2012; Eilert & Nappier Cherup, 2020). Most studies in this area have concentrated on large corporations and their involvement in public debates surrounding social and environmental issues (García-Rosell, 2023; García-Rosell & Haapakoski, 2023). Particular attention has been devoted to causes such as climate change, immigration, land rights, gender equality and LGBTQ+ rights, among others. While these areas warrant urgent scholarly attention for their potential to drive social, political and legislative change, comparatively little research has examined animal activism and advocacy within business organisations.

For instance, corporate activism, brand activism and CEO activism have become well-established management concepts used to describe the deliberate engagement of business actors in public debates concerning controversial sociopolitical issues (Eilert & Nappier Cherup, 2020; Simon, 2011). While this form of activism may align with the principles of corporate social responsibility, it differs in its progressive orientation and its potential to provoke public controversy and pose significant risks to firm value (Branicki et al., 2021). Indeed, when companies adopt a political stance, they may attract support from certain stakeholders while risking alienation from others. As a result, some organisations may engage only in symbolic gestures towards addressing social problems and political claims, while neutralising contentious issues through compromise to avoid alienating potential consumers (Simon, 2011). Corporate and CEO activism also differs fundamentally from traditional "bottom-up" social activism, which is typically led by NGOs and citizens with direct, often adverse, experiences of the issues they advocate for (Branicki et al., 2021). Activism within organisational contexts is driven by CEOs' personal values, stakeholder expectations and, at times, instrumental goals linked to business strategy. For responsible managers, this raises a critical question: If companies can take political stances on human rights and environmental issues, why not on the exploitation of animals embedded within their own business models?

Corporate and CEO activism emerges from situations where values are in conflict, demanding change across industries and social institutions. Although limited scholarly attention has been given to animal-related causes, several corporations, such as The Body Shop, Lush, Patagonia and The North Face, have publicly advocated for animal rights and welfare. In addition, some studies have examined small businesses engaging politically to improve the welfare of working animals (e.g. García-Rosell, 2023; García-Rosell & Haapakoski, 2023). However, whether grounded in a rights-based or welfarist perspective, this form of corporate and CEO activism often focuses on campaigning against specific industry practices and advocating for single species – mostly vertebrates – without necessarily transforming entire business models. To address these limitations, the following section introduces an interspecies approach to activism within organisations.

An interspecies approach to company activism

Drawing on the case of Transfarmation Austria and an interview with Doris Schneeberger, we elaborate on the premises underlying an interspecies approach to company activism. While the terms corporate and CEO activism typically refer to the political engagement of large firms, we adopt the term company activism to encompass organisations of varying sizes. To align company activism more closely with an interspecies approach, we propose three guiding principles: anti-speciesism, interventionism and transformation.

Anti-speciesism principle

By focusing on a single species and emphasising human interests, business organisations engaging in animal advocacy may inadvertently reinforce speciesism, whereby certain animals are deemed more worthy of protection based on their perceived capacities or aesthetic appeal, while others – considered of lower status – receive less attention or concern (see Horta, 2010; Schneeberger, 2023). Through its "Häagen-Dazs Loves Honey Bees" initiative, for instance, Nestlé advocates for the health of bees and promotes bee-friendly farming practices. Nevertheless, it is difficult to find comparable examples of companies taking a political stance in support of mosquito protection – despite their vital role as pollinators – because these insects are predominantly framed as pests or disease vectors to be eradicated (see Davies & Sayers, 2023; Valtonen et al., 2020). This asymmetry reveals the deep species hierarchies embedded in corporate animal activism.

Considering this limitation, this interspecies approach is based on an anti-speciesism principle. As a philosophical stance, anti-speciesism challenges the hierarchical distinction that humans draw between themselves and other animal species. In doing so, anti-speciesism ascribes moral consideration to all sentient animals without exception (Singer, 1975). The term aligns with critiques of other systems of discrimination, such as racism, sexism and classism, by arguing that discrimination based on species membership is morally problematic. Anti-speciesism calls for recognising animals as sentient beings with intrinsic value, rejecting the assumption that human interests and needs are more important than those of other species. From this perspective, practices such as industrial farming, animal experimentation and the exploitation of animals in entertainment or tourism are seen as forms of structural violence maintained by anthropocentric norms. By positioning humans as the central source of meaning and value, anthropocentrism provides the worldview that legitimises speciesism (Singer, 1975).

In the context of company activism, an interspecies approach invites a critical examination of how advocacy for animal rights and welfare reinforces species hierarchies. An interspecies lens challenges this anthropocentric bias by questioning the legitimacy of organisational practices across species boundaries. It encourages companies to move beyond tokenistic or species-specific campaigns and instead question the broader ethical implications of their operations for all sentient beings. This includes examining supply chains, labour conditions, environmental impacts and technological interventions not only in terms of human wellbeing but also in relation to the lives and habitats of other species. This perspective highlights the relational entanglement of different species within organisations. For example, the exploitation of animals in industrial agriculture, as reflected in the case of Transfarmation Austria, is bound up with organisational practices that affect human communities, farmed animals, wild species (vertebrates and invertebrates), ecosystems and nature. By recognising these entanglements, an interspecies approach seeks to dismantle systems of domination and inequality that span across different species. An interspecies approach is not merely about extending rights to animals; it is about re-evaluating the ethical foundations of organisational and social behaviour in general.

Transformation principle

The case discussed in this chapter, Transfarmation Austria, highlights the importance of transforming existing organisational models. Transfarmation – the conversion of farms using animals to animal-free farms and/or vegan sanctuaries – offers an opportunity to build bridges between humans and other species. Rather than deepening social polarisation, the principle of transformation seeks to promote actions that benefit both animals and humans, who are embedded within the same systems of exploitation. At times, animal activism focuses primarily on taking political stances without directly engaging in systemic change. While advocating for animal rights remains crucial, the lack of practical solutions risks trapping the debate in ideological polarisation, preventing meaningful systemic change. For instance, Oatly has become well known for taking an explicit political stance against the dairy industry. By promoting plant-based alternatives, the company implicitly draws attention to the animal welfare issues inherent in traditional dairy farming, including the instrumental use of animals and their living conditions (Koch, 2020). Through its marketing campaigns – such as using packaging to portray dairy production as unethical and environmentally destructive – Oatly positions itself in confrontation with conventional dairy farming. Over time, the company adopted a more proactive approach by collaborating with farmers willing to transform their production models towards more sustainable, plant-based agricultural practices (Koch, 2020).

A nucleus of change for animals is Hof Narr, a small-scale farm in Switzerland that has been converted into a sanctuary for animals rescued from suffering or slaughter. On this sanctuary and vegan farm, animals are allowed to live in conditions appropriate to their species, free from exploitation. Hof Narr offers holiday camps and educational programmes – particularly for children and school groups – designed to foster empathy towards animals and raise awareness about the realities of industrial farming (see Köllen & Schneeberger, 2023; Schneeberger, 2026). By encouraging direct encounters with animals as sentient individuals rather than production units, these activities aim to reshape participants' understanding of human–animal relations and promote more ethical forms of coexistence. The founder of the sanctuary, Sarah Heiligtag, extends this transformative approach beyond Hof Narr through her initiative Transfarmation Switzerland, which supports other farmers in developing alternative business models that no longer rely on animal exploitation. By helping farmers design economically viable, animal-friendly enterprises, Heiligtag's activism bridges the gap between ethical commitment and practical transformation, illustrating how agricultural practices can evolve towards more compassionate and sustainable systems (Köllen & Schneeberger, 2023).

These examples show that an interspecies approach to company activism is not about deepening polarisation or confrontation in society, but about fostering cooperation across sectors to enable meaningful transformation of established business models. As Köllen and Schneeberger (2023) note, converting an intensive animal-feeding operation into a zero-suffering farm is a complex process that demands long-term commitment and societal support. In this context, anti-speciesism becomes a precondition for Transfarmation, as it underpins activism oriented towards solidarity, care and the creation of systems where both humans and other species can coexist without suffering or exploitation. Farms provide an especially interesting context, as Doris argues in the interview for this chapter. Farmers have the potential to change because of their close and continuous relationships with animals. Such interspecies bonds can reshape individual mindsets and, ultimately, transform the entire business. However, transformation does not occur spontaneously – it requires intervention, the next principle we discuss.

Interventionist principle

Interventionism is characterised by a problem-solving orientation, direct engagement with the complexities of real organisational settings, collaborative involvement of multiple stakeholders and a commitment to producing tangible, practice-oriented outcomes (Baard & Dumay, 2020). Building on the definition of interventionist research (Dumay & Baard, 2017), the interventionist principle guiding an interspecies approach to company activism entails the active engagement of an interventionist – whether a researcher or consultant – with organisational actors to address concrete social problems. This principle underscores a relational process of organisational change, facilitated through training, workshops and programmes that challenge and expand participants' moral circles to include other species (Schneeberger, 2023). Drawing on the idea of implementing change to assess its practical impact, Transfarmation Switzerland exemplifies an interventionist initiative that has already supported the transformation of over 150 farms. These farms not only serve as concrete examples of business model transformation but also act as advocates for broader systemic change towards zero-suffering farming practices (Köllen & Schneeberger, 2023) and post-lethal agriculture.

Transfarmation Austria conducts similar work under the leadership of Doris, who, as an animal rights activist and interventionist consultant, drives transformation by introducing an anti-speciesist perspective. To facilitate change, she integrates Kurt Lewin's (1947) theory of change with Anna Freud's (1984) theory of defence mechanisms in her interventionist approach (Schneeberger, 2023). Lewin's three-stage model – Unfreeze, Change and Refreeze – conceptualises change as a process of preparing an organisation for transformation, implementing new practices and stabilising them to ensure their sustainability. Freud's theory, on the other hand, suggests that the ego employs unconscious defence mechanisms to manage anxiety and internal conflict, processes that can both inhibit and enable transformation. From this perspective, humans tend to resist "unfreezing" because they seek to preserve the stability of their existing mental and moral systems. However, as Doris argues in relation to advocacy for animals and the reduction of speciesism, witnessing the violence and suffering inherent in animal husbandry can produce a moral shock – an empathic and emotional reaction that may ultimately catalyse change and the acceptance of a less speciesist moral system (Schneeberger, 2023). The success of such an intervention depends on the full immersion of the interventionist researcher or consultant in the organisational culture, engaging as closely as possible with the phenomenon and the actors involved (see Baard & Dumay, 2020).

Interventionism concerns not only organisational but also social change. While organisational change involves transforming culture, relationships, strategies and job designs within companies, social change focuses on enhancing the capacity of organisational actors to develop as competent individuals and citizens aware of social values and power relations (see Baard & Dumay, 2020). Lewin's (1947) work on social change and emancipation is often associated with marginalised groups – such as the poor, ethnic minorities, religious groups and women – seeking freedom and autonomy from oppression. Yet, emancipation is also relevant in organisational life, where other animal species can be seen as seeking liberation from violence, suffering and injustice in exploitative human systems (see Schneeberger, 2023). The case of Transfarmation Austria illustrates how interventionist-driven activism can challenge speciesism and promote the transformation necessary for a more humane society and institutions. From this perspective, an interspecies approach to company activism aims to cultivate anti-speciesist awareness and empathy towards other species. By fostering innovation and developing practical solutions, Transfarmation envisions a future in which farming is not only economically viable but also aligned with ecological sustainability and compassion for animals.

Directions for future research

An interspecies approach to company activism offers an opportunity for advancing the literature on activism within organisational contexts. One compelling direction for future research is to critically interrogate the politics of species representation in company activism. As illustrated by the contrast between widespread advocacy for bees and the lack of similar efforts for mosquitoes – or the predominant focus on vertebrates over invertebrates in animal advocacy – companies tend to selectively engage with species based on cultural symbolism, perceived utility and public sentiment (see Davies & Sayers, 2023). This selective advocacy reflects entrenched species hierarchies and anthropocentric biases, shaping which animals are deemed worthy of protection and which are excluded from ethical consideration. To address these dynamics, future studies could examine how social discourses construct species representations and how these constructions shape company activism. Comparative case studies across industries such as agriculture, fashion and tourism could reveal patterns of inclusion and exclusion, shedding light on the ethical and strategic choices companies make when engaging with animal-related issues. Such research would offer valuable insights into how an interspecies approach to company activism might trigger more inclusive and holistic models – ones that challenge speciesist assumptions and promote ethical transformation across species boundaries.

Another fruitful avenue for research lies in exploring organisational and social change through an interspecies approach to company activism in other economic sectors. Building on examples like Hof Narr and Transfarmation Austria, scholars could investigate how direct human–animal relationships – particularly those formed in an organisational setting – can catalyse ethical shifts in business models and consumer behaviour (see Köllen & Schneeberger, 2023). Research could focus on the mechanisms and practices through which such transformations occur, including the role of training programmes, participatory workshops and stakeholder engagement in expanding the moral circle to include other species (see Delmestri & Schneeberger, 2022; Schneeberger, 2023, 2025). The interventionist principle, which emphasises collaborative problem-solving and tangible outcomes, is particularly well-suited to this task. It allows researchers to co-create solutions with organisational actors, test new models of interspecies cooperation and document the challenges and successes of implementing zero-suffering business models.

As an interspecies approach to company activism seeks to dismantle systems of domination and inequality that extend across species boundaries, future research could critically examine supply chains, labour conditions, environmental impacts and technological interventions – not only in terms of human wellbeing but also in relation to the lives and habitats of other species. By acknowledging these interconnections, scholars can advance a more holistic understanding of corporate responsibility – one that integrates ecological integrity alongside the wellbeing of both humans and other species. Such research is essential for informing policy development and guiding companies towards ethical standards that reflect the interconnectedness of all life forms (see Schneeberger & Traavik, 2025). In sum, by engaging with these themes, future research can contribute to building more compassionate and sustainable business models that challenge anthropocentric norms and speciesism and promote practical interventions towards systemic change in business and society.

Ethical implications

Adopting an interspecies approach to company activism raises important ethical considerations, as it challenges the anthropocentric foundations of contemporary business ethics and activism in organisations. It requires companies to expand moral consideration beyond humans and to acknowledge the intrinsic value and vulnerability of nonhuman life. This perspective compels organisations to question practices that normalise animal exploitation within production and consumption systems. Ethically, it calls for a redefinition of business responsibility – one that incorporates duties towards nonhuman stakeholders and recognises their interests as morally relevant (Schneeberger & Karimova, 2023). However, this expansion also introduces dilemmas: balancing competing needs among species, assessing the moral legitimacy of animal use in specific contexts and reconciling economic imperatives with ecological and ethical obligations.

While emotionally powerful content such as documentaries depicting the killing or exploitation of animals can be effective in challenging speciesism and provoking moral reflection, it also raises concerns about psychological harm, informed consent and the instrumentalisation of suffering (see Schneeberger, 2023). Although the goal of an interspecies approach to company activism is to foster transformation through awareness and engagement, this must be carefully balanced with responsible communication practices. Graphic portrayals of violence may alienate audiences, cause distress or reinforce the image of animals as passive victims rather than sentient beings with agency. Therefore, companies and activists must carefully consider how such content is framed, ensuring it is contextualised within educational or transformative narratives and accompanied by opportunities for reflection and dialogue. The material used in activism should aim not only to disrupt harmful norms but also to cultivate empathy, solidarity and a deeper understanding of multispecies entanglements.

An often overlooked but ethically significant aspect of any form of company activism is the wellbeing of the researcher, activist or consultant who leads organisational interventions. As Doris reflected in her interview, many activists begin their work with deep passion and conviction, only to later encounter the emotional and psychological toll of striving for meaningful change in resistant or slow-moving systems. The challenge of persuading others, confronting well-established norms and witnessing ongoing suffering can lead to frustration, burnout and eventual disengagement. Ethically, this raises important questions about how organisations and research communities can support the emotional and physical wellbeing of those engaged in activism – through peer networks, reflective practices and institutional recognition of the emotional work involved in interspecies advocacy.

Practical implications for responsible management

Adopting an interspecies approach to company activism carries several practical implications for responsible management, reshaping how organisations define their ethical responsibilities, engage stakeholders and develop sustainability strategies. First, it expands companies' political and ethical responsibilities beyond human stakeholders to include animals and ecosystems affected by organisational activities (Schneeberger & Karimova, 2023). This shift challenges anthropocentric assumptions and calls for a broader understanding of political accountability – one that recognises the interconnectedness of human and nonhuman lives in business and society. Second, it calls for a reorientation of organisational values and culture. An interspecies approach encourages companies to embed principles of care, empathy and multispecies justice into their activism and internal practices. This may involve creating spaces for ethical reflection, supporting employee-led initiatives and fostering interspecies awareness through education, training and direct engagement with other species. These efforts help cultivate a culture of ethical sensitivity and solidarity across species lines.

Third, this implies a commitment to activism that reimagines business and triggers radical societal change. Companies must not just take stances but actively seek alternatives to exploitative practices – such as transitioning from animal-based to plant-based production models, investing in cruelty-free technologies and redesigning products and services to minimise harm to other species (see Köllen & Schneeberger, 2023). This transformation is not only ethically necessary but also strategically relevant, aligning with the growing consumer demand for greater sustainability, transparency and business models that support the wellbeing of both humans and nonhumans. Fourth, an interspecies approach requires collaborative engagement across disciplines and sectors. Companies must work jointly with researchers, activists and civil society actors to co-create solutions that are both economically viable and ethically sound. Understanding the specific species involved and their entanglements with organisational practices demands interdisciplinary collaboration – drawing on fields such as animal studies, environmental ethics, sociology, psychology and business.

Finally, companies engaging in interspecies activism must address the wellbeing of those who drive transformation, including researchers, consultants and activists. The emotional labour of advocating for systemic change – often in the face of resistance or slow progress – can lead to burnout and disappointment. Supporting the emotional resilience of these individuals, recognising their contributions and fostering a culture of care within organisations are essential for sustaining long-term activism. Ethical management must therefore include mechanisms for reflection, peer support and institutional recognition of the moral commitment required to build a more just and compassionate society for both humans and nonhumans. Seen in this light, company activism is not a branding strategy but an ethical commitment that shapes how organisations engage with multispecies responsibility.

Reflective activities inside/outside the classroom

A Revision questions that test knowledge ("what?")

- A1 According to the chapter, what concepts have been used to describe activism in an organisational context and what limitations have been identified?
- A2 What are the two dominant ethical perspectives associated with animal activism in organisations?
- A3 How does the case of Transfarmation Austria challenge prevailing understandings of activism within organisational settings?
- A4 What is meant by an interspecies approach to company activism and how does it differ from traditional forms of activism in organisations?

B Reflective questions that build understanding ("so what?")

- B1 Why is it important for an interspecies approach to company activism to be guided by the principles of anti-speciesism, transformation and intervention?

- B2 How can company activism be designed to support both human wellbeing and the wellbeing of other species?
- B3 Why is it essential to question speciesist and anthropocentric assumptions in company activism?
- B4 Why should all species be considered potential beneficiaries of company activism?

C Take action ("do")

- C1 What aspects should I consider before engaging in activism from an interspecies perspective?
- C2 How can I raise awareness among peers or colleagues about the importance of anti-speciesism and the need to transform business models towards zero-suffering?
- C3 How could an interspecies approach to company activism be applied within my university or workplace?
- C4 How might I design an activist initiative that benefits both humans and nonhuman species?

D Ideas to discuss with colleagues, friends and family ("share")

- D1 To what extent is it morally acceptable to use documentaries portraying violence and animal suffering to raise awareness in activist campaigns?
- D2 Can activism that advocates for the better treatment of animals be considered interspecies activism? Why or why not?
- D3 Are there examples in your field of work where an organisation has taken a political stance that aligns with the interspecies approach to company activism discussed in this chapter?
- D4 Have you ever reflected on company campaigns advocating for insects or other invertebrates? What are your thoughts on such initiatives?

Key ideas

Interspecies perspective on company activism

An interspecies approach to company activism extends the scope of corporate activism and CEO activism by including all species within the moral circle of business political activity. It refers to a company's commitment to advocating for anti-speciesism – the rejection of species-based discrimination – and actively working towards transforming business models that reduce or eliminate suffering and exploitation across species lines (Schneeberger, 2023; Singer, 1975).

Speciesism

Speciesism is the unjustified, disadvantageous consideration or treatment of those who are not classified as belonging to one or more particular species (Horta, 2010).

Anti-speciesism

Anti-speciesism is an ethical stance that rejects discrimination based on species membership. It holds that all sentient beings – human and nonhuman alike – deserve equal moral consideration, particularly regarding their capacity to suffer. Anti-speciesism challenges the assumption that human interests inherently outweigh those of other species and calls for the inclusion of animals within the scope of justice, care and political concern (Singer, 1975).

Transfarmation

Transfarmation refers to the structured transition from animal-based farming systems to plant-based or sanctuary-oriented models that eliminate exploitation. It emphasises practical change over symbolic protest, combining ethical commitment with economically viable alternatives that reduce suffering across species (Köllen & Schneeberger, 2023).

Interventionism

Interventionism involves active engagement with organisations to co-create ethical transformation. Rather than observing from a distance, interventionists collaborate with managers and stakeholders to implement change, expand moral consideration and develop practical alternatives to exploitative systems (Dumay & Baard, 2017; Schneeberger, 2023).

References

Aronczyk, M. (2013). Market(ing) activism: Lush cosmetics, ethical oil and the self-mediation of protest. *JOMEC Journal*, 3. doi:10.18573/j.2013.10256.

Baard, V. C., & Dumay, J. (2020). Interventionist research in accounting: Reflections on the good, the bad and the ugly. *Accounting & Finance*, 60(3), 1979–2006. doi:10.1111/acfi.12409.

Beers, D. L. (2006). *For the prevention of cruelty: The history and legacy of animal rights activism in the United States*. Swallow Press / Ohio University Press.

Branicki, L., Brammer, S., Pullen, A., & Rhodes, C. (2021). The morality of "new" CEO activism. *Journal of Business Ethics*, 170, 269–285. doi:10.1007/s10551-020-04656-5.

Briscoe, F., & Gupta, A. (2016). Social activism in and around organisations. *The Academy of Management Annals*, 10(1), 671–727. doi:10.5465/19416520.2016.1153261.

Corvellec, H., & Stål, S. (2019). Qualification as corporate activism: How Swedish apparel retailers attach circular fashion qualities to take-back systems. *Scandinavian Journal of Management*, 35(4), 101046. doi:10.1016/j.scaman.2019.03.002.
Curtin, N., & McGarty, C. (2016). Expanding on psychological theories of engagement to understand activism in context(s). *Journal of Social Issues*, 72, 227–241. doi:10.1111/josi.12164.
Davies, O., & Sayers, T. D. J. (2023). Honeybee bias and bee-washing: Effects of vertebrate-centric care? In L. Tallberg & L. Hamilton (Eds), *The Oxford handbook of animal organisation studies* (pp. 313–330). Oxford University Press. doi:10.1093/oxfordhb/9780192848185.013.21.
de Bakker, F. (2012). Exploring network of activism on corporate social responsibility: Suggestions for a research agenda. *Creativity and Innovation Management*, 21(2), 212–223. doi:10.1111/j.1467-8691.2012.00641.x.
Delmestri, G., & Schneeberger, D. (2022). Organisational change and work spirituality: Expanding the moral circle. In Y. Altman, J. Neal & W. Mayrhofer (Eds), *Workplace spirituality: Making a difference* (pp. 95–110). De Gruyter. doi:10.1515/9783110711349-006.
Dumay, J., & Baard, V. (2017). An introduction to interventionist research in accounting. In Z. Hoque, L. D. Parker, M. Covaleski & K. Haynes (Eds), *The Routledge companion to qualitative accounting research methods* (pp. 265–283). Routledge.
Eilert, M., & Nappier Cherup, A. (2020). The activist company: Examining a company's pursuit of societal change through corporate activism using an institutional theoretical lens. *Journal of Public Policy & Marketing*, 39(4), 461–476. doi:10.1177/0743915620947408.
Fennell, D. A. (2014). Exploring the boundaries of a new moral order for tourism's global code of ethics: An opinion piece on the position of animals in the tourism industry. *Journal of Sustainable Tourism*, 22(7), 983–996. doi:10.1080/09669582.2014.918137.
Freud, A. (1984). *Das Ich und die Abwehrmechanismen*. Fischer.
García-Rosell, J.-C. (2023). Husky kennels as animal welfare activists: Multispecies relationships as drivers of institutional change. In In L. Tallberg & L. Hamilton (Eds), *The Oxford handbook of animal organisation studies* (pp. 397–408). Oxford University Press. doi:10.1093/oxfordhb/9780192848185.013.26.
García-Rosell, J.-C., & Haapakoski, A.-E. (2023). Political corporate social responsibility in small tourism business: The case of a Finnish sled dog kennel. *CABI Tourism Cases*. doi:10.1079/tourism.2023.0009.
Hartman, C. L., & Beck-Dudley, C. L. (1999). Marketing strategies and the search for virtue: A case analysis of the Body Shop International. *Journal of Business Ethics*, 20, 249–263. doi:10.1023/A:1006057807115.
Horta, O. (2010). What is speciesism? *The Journal of Agricultural and Environmental Ethics*, 23, 243–266. doi:10.1007/s10806-009-9205-2.
Koch, C. H. (2020). Brands as activists: The Oatly case. *Journal of Brand Management*, 27, 593–606. doi:10.1057/s41262-020-00199-2.
Köllen, T., & Schneeberger, D. (2023). Avoiding unnecessary suffering: Towards a moral minimum standard for humans' responsibility for animal welfare. *Business Ethics, the Environment & Responsibility*, 32, 1139–1149. doi:10.1111/beer.12565.
Lewin, K. (1947). Frontiers in group dynamics: Concept, method and reality; social equilibria and social change. *Human Relations*, 1(1), 5–41. doi:10.1177/001872674700100103.

Rodriguez García, M. (2024). Platero y yo. La visión juanramoniana de la fauna en la literatura animalística española. *Revista de la Facultad de Filosofía y Letras, 74*, 619–654.

Saunders, C. (2022). Activism. In D. A. Snow, D. Porta, B. Klandermans & D. McAdam (Eds), *The Wiley-Blackwell encyclopedia of social and political movements*. Wiley. doi:10.1002/9780470674871.wbespm002.pub2.

Schneeberger, D. (2023). How can we reduce speciesism? A psychological approach to a social problem. In L. Tallberg & L. Hamilton (Eds), *The Oxford handbook of animal organisation studies* (pp. 241–257). Oxford University Press. doi:10.1093/oxfordhb/9780192848185.013.16.

Schneeberger, D. (2024). *Envisioning a better future for nonhuman animals: Towards future animal rights declarations*. Palgrave Macmillan. doi:10.1007/978-3-031-75332-9.

Schneeberger, D. (2025). "Expanding the moral circle" – Zum Einsatz audiovisueller Medien zur Förderung von Antispeziesismus und Antisubstratismus in Bildungskontexten. *TIERethik – Zeitschrift zur Mensch-Tier-Beziehung*, 17(1), 141–165. doi:10.58848/tierethik.2025.1.141.

Schneeberger, D. (2026). Nonhuman animals in organizations: Products, slaves, or workers? In R. Bendl, C. Khan & A. Schmidt (Eds), *Frauen die Wissen schaffen (FDWS) – Buchprojekt zu 35 Jahre Frauenarbeit an der Wirtschaftsuniversität Wien*. Springer.

Schneeberger, D., & Karimova, G.-S. (2023). Representing invisible stakeholders in the political democratic discourse: The case of climate and animal rights activism. In J. Fritz & N. Tomaschek (Eds), *Partizipation: Das Zusammenwirken der Vielen für Umwelt, Wirtschaft und Demokratie* (pp. 79–92). Waxmann.

Schneeberger, D., & Traavik, L. (2025). How speciesist is AI? – discussing AI, speciesism and vegaphobia with ChatGPT – Implications for inclusive workplaces. In O. Kyriakidou & J. Vassilopoulo (Eds), *AI and diversity in a datafied world of work: Will the future of work be inclusive?* (pp. 225–247). Emerald Publishing. doi:10.1108/S2051-233320250000012025.

Simon, B. (2011). Not going to Starbucks: Boycotts and the out-scouring of politics in the branded world. *Journal of Consumer Culture*, 11(2), 145–167. doi:10.1177/1469540511402448.

Singer, P. (1975). *Animal liberation: A new ethics for our treatment of animals*. New York Review.

Tallberg, L., Välikangas, L., & Hamilton, L. (2022). Animal activism in the business school: Using fierce compassion for teaching critical and positive perspectives. *Management Learning*, 53(1), 55–75. doi:10.1177/13505076211044612.

Valtonen, A., Salmela, T., & Rantala, O. (2020). Living with mosquitoes. *Annals of Tourism Research*, 83, 102945. doi:10.1016/j.annals.2020.102945.

Youngs, R. (2019). *Civic activism unleashed: New hope or false dawn for democracy?* Oxford University Press. doi:10.1093/oso/9780190931704.001.0001.

Żuchowska-Skiba, D. (2026). Activism. In J. Marques (Ed.), *Encyclopedia of diversity, equity, inclusion and spirituality*. Springer. doi:10.1007/978-3-031-76750-0_49.

PART V

Reflections and implications for the future

13

PEDAGOGICAL IMPLICATIONS

Introduction

This chapter explores the pedagogical implications of expanding management concepts and the discussion on responsibility in organisations beyond human needs to include the wellbeing of nonhuman species. Building on the central claim that an interspecies perspective on responsible management is an ethical imperative, we consider how the insights presented in this book can inform teaching practices, curriculum design and educational policy.

Across the preceding chapters, animals have appeared not as background actors but as workers, consumers, innovators and political subjects in their own right. To continue marginalising them within business education would simply reproduce the anthropocentrism this volume has sought to unsettle. This chapter therefore moves from diagnosis to application. It asks not simply how animals can be added to existing curricula, but how business education itself must evolve if it is to remain intellectually credible and ethically responsive in a multispecies world.

An interspecies perspective on responsible management can be put into practice in different ways in an educational setting. Rather than offering strict guidelines on how to teach and develop learning activities, the book is an invitation for educators and students to reflect on the role that animals play in relation to the concepts included in the business school curriculum. It makes visible how nonhuman species have been excluded from business education or only included as resources to be used and exploited for the sake of human needs. The chapters in this book encourage readers to think more critically about how responsibility is understood and framed in business education. Can management and marketing approaches be called responsible or ethical if they fail to include animals and other nonhuman

DOI: 10.4324/9781003585541-19

species? Can sustainability be achieved without considering the needs and interests of nonhuman species? Critically reflecting on these questions can lead us to new ways of thinking and acting with empathy, compassion and respect towards humans and nonhuman species living on this planet.

What does it mean to teach and learn in light of the perspectives developed in this book? This chapter addresses that question in two stages. We begin by considering how the book can be used in practice. We then turn to a set of broader pedagogical considerations that follow from an interspecies approach to responsible management education.

Using this book in practice

This book is designed to function as a complete course in Animals and Business, or Animals and Responsible Management. Taken together, the chapters provide the conceptual foundations, empirical illustrations, critical debates and pedagogical prompts necessary to structure a full semester module for undergraduate or postgraduate students. We recognise that many colleagues will be encountering animal-focused scholarship for the first time and may effectively be learning alongside their students. For that reason, each chapter brings together key ideas, examples and suggested exercises so that the intellectual and practical resources required to run a course are contained within the book itself.

At the same time, the structure is intentionally modular. Individual chapters can be incorporated into existing courses in marketing, accounting, strategy, tourism, sustainability, governance or organisational studies. In such contexts, a chapter can operate as a ready-made session, introducing core concepts, framing discussion questions and inviting students to experiment with applying an interspecies lens to familiar business topics. Colleagues are, of course, free to extend or supplement the material, but they need not design new content themselves unless they wish to do so.

Whether used as the foundation for a dedicated module or integrated into mainstream curricula, the purpose of the book remains the same: That is, to equip educators with the conceptual tools, empirical insights and pedagogical resources needed to embed animals and other nonhuman species more systematically, confidently and creatively within business education.

For students who have worked through the preceding chapters, this final section is not a new beginning but a reframing of what you have already encountered. By this point in the book, business can no longer appear quite as it did at the beginning. You have seen animals emerge as workers, commodities, co-creators of value, political subjects and institutional absences. The task now is to carry those insights back into familiar business concepts and practices. You are not expected to have prior expertise in animal studies; rather, you are invited to reflect on how your understanding of business may already have shifted. If certain assumptions now feel unsettled – about value, responsibility, labour or sustainability – that signals that learning has taken place.

Building on these practical considerations, we now outline four pedagogical dimensions that can guide the development of interspecies approaches in business education: Learning as an ethical and imaginative exercise; critical reflexivity; learning with animals; and reassessing the business curriculum.

Learning as an ethical and imaginative exercise

Business education has been characterised by reinforcing anthropocentric narratives, presenting sustainability as a means to improve efficiency rather than challenge ethical foundations (Kopnina et al., 2025). Similarly, the idea of responsible business is discussed in relation to human stakeholders and the natural environment, which often considers animals part of nature. An interspecies approach challenges this by highlighting the agency, sentience and intrinsic value of animals. Rather than viewing supply chains solely through the lens of carbon or cost, students can examine how industries like meat and fashion depend on animal suffering and explore alternatives such as plant-based systems, ethical sourcing or degrowth to reduce animal exploitation. This enables teachers and learners to reimagine sustainability not as a technical fix, but as a transformative practice grounded in care and responsibility (Kopnina, 2020).

An interspecies perspective, then, is not just about adding animals to sustainability teaching and learning. Rather, it is about cultivating critical thinking, empathy, compassion and responsibility in ways that will be useful to students elsewhere in their studies as well as their professional and personal lives (see Tallberg et al., 2022). It creates space for dialogue, complexity and disagreement as part of building more inclusive, just futures for humans and nonhuman species (Wadham et al., 2023). From this perspective, responsible business education becomes less about transmitting expertise and more about practising moral imagination and relational accountability across species. As an ethical and imaginative exercise, it encourages educators and students to critically examine the role animals play in business and society through multiple ethical lenses. The analysis centres on the dignity and moral consideration of individual animals, rather than on species collectively or nature as the focus of analysis (see Fennell & Sheppard, 2021).

Critical reflexivity

The book also calls for moving beyond critical thinking to include pedagogies that foster critical reflexivity. While critical thinking refers to our ability to adopt a disciplined approach to problem-solving by relying on existing knowledge, critical reflexivity alludes to the social process of questioning assumptions embedded in both theory and professional practice (see

Cunliffe, 2002, 2004; García-Rosell, 2012). The cases discussed in the book – such as the animals working in tourism and public health – can help students reflect on interspecies interdependence, power relations, care and informed consent, while also fostering critical reflexivity about their own assumptions regarding the status of animals in society and their moral obligation towards them. Bringing animals into responsible management teaching and learning encourages teachers and students to think differently about fairness, justice, inclusion and exclusion. Instead of seeing animals as irrelevant to business or organisation studies, this approach asks challenging questions: Who is granted stakeholder status and who is not? Whose work and services are ignored? Who keeps our systems running but never gets credit? And whose voices are neglected or regarded as not worthy of attention?

By promoting critical reflexivity in the business classroom, we go beyond simply adding in more diverse groups to our analysis and instead focus on structural critiques of the deeper systems and organisational practices that create inequality and injustice. It helps us recognise how animals are marginalised alongside humans, especially through connections to gender, social class, race and rural life. In doing so, it prepares students to think about fairness and justice that goes beyond humans – important in areas like tourism, agriculture, sustainability, policy and planning – expanding their ethical considerations and awareness. Even if judgments about what is right and just are often treated as subjective, suffering is not. Critical reflexivity on taken-for-granted organisational practices that affect nonhuman species therefore has broader relevance for understanding the interconnectedness of humans and animals (Tallberg et al., 2022).

An interspecies lens further supports reconsidering responsibility and governance – what we might call responsive governance. Drawing on the political ethics of care (Tronto, 2020), both teachers and students are invited to consider who benefits and who bears the costs within organisational systems depending on animals and to critically analyse how practices of care and control converge under conditions of economic pressure and biological vulnerability. More broadly, this pedagogical shift reframes learning as the cultivation of responsiveness, recognising that the capacity to perceive and interpret animals' signals while acknowledging that knowledge is always partial, situated and co-produced. Here, education becomes less about transmitting expertise and more about practising reflexivity, moral imagination and relational accountability across species. It involves being critically reflexive about controversial business practices as an opportunity to inspire courage, empathy and action for change towards more ethical and responsible business practices (see Tallberg et al., 2022).

Learning with animals

Teaching from an interspecies perspective means rethinking learning itself as an interspecies relational process. This shift in mindset calls for a new way of teaching – one that values different forms of knowledge and fairness in whose experiences we learn from. In terms of pedagogical approaches, it requires moving away from lecture-based instruction about nonhuman species towards embodied, experiential and place-based learning with animals and other nonhuman species (see Höckert et al., 2019). It starts with the understanding that animals communicate through presence, expressing themselves in how they act, resist and relate. It calls for attentiveness to what animals can teach us about empathy, vulnerability and coexistence and that learning can happen through co-presence, observation and awareness of other ways of being in the world. To better adapt to this pedagogical context, business educators must be more innovative in their pedagogical methods, using fieldwork, study trips, careful embodied observation, ethnographic storytelling, documentaries, reflective journaling and participatory mapping to help students grasp interspecies entanglements. Through these pedagogies, business students will begin to view animals not as background props but as active participants in social and economic life.

From this perspective, learning with animals opens opportunities to create new knowledge and ways of looking into theoretical concepts. For example, recognising animals as workers invites educators and students to reflect on how labour, value and responsibility are distributed across species and what this means for the future of work and leadership. Thinking about animals as workers encourages learners to consider how compassion, welfare and sustainability can shape leadership beyond standard models of efficiency or productivity. By reflecting on who counts as a worker, what counts as labour and how workplace ethics are expanded beyond humans, learning with animals can be regarded as an opportunity to foster more inclusive, relational modes of thinking. In a similar way, the learning experience with chickens as sentient beings embedded within socio-metabolic systems can encourage students to interrogate the production of invisibility itself (Colombino & Bruckner, 2023). It enables them to see how cultural norms, technological infrastructures, welfare labels and policy frameworks collectively obscure the labour, lives and ecological consequences underpinning everyday consumption of animals.

A pedagogical approach that emphasises learning with animals must take into account the educational and cultural context in which it is implemented. Much of the literature on responsible and sustainable business education assumes small, homogeneous classrooms (Wadham et al., 2023). In practice, as we know from teaching in large, diverse universities, this is rarely the case. For instance, many European universities host students from diverse European and non-European backgrounds. At the same time, some non-European students may be motivated to study at "Western" universities to better understand capitalism and succeed within that system, rather than to critique and question it. These dynamics may create resistance to interspecies perspectives and to the idea of learning with animals. However, this diversity is also a pedagogical strength. Many students bring lived experiences of multispecies entanglement – whether from agricultural or urban settings in countries such as Nigeria, India and Colombia, among others – where humans and animals share and navigate space on a daily basis. Drawing on these experiences with animals in different cultural and social contexts allows us to meet students (and teachers) where they are while also encouraging critical ethical reflection.

Reassessing the business curriculum

Bringing an interspecies perspective into teaching about business concepts invites us to rethink what business education is for. Instead of treating managerial and organisational skills as neutral, this book should help students see them as a way of shaping the world and our social reality – deciding what matters, what counts and what gets ignored. For business schools, this book is a call to move beyond abstract debates about sustainability or ethics and to help students see that business decisions have moral and political implications for both humans and nonhuman species. Rather than passively learning about business texts or practices, students are prompted to think about how animals affect and are affected by those practices and how their treatment reflects broader cultural and ecological relationships. From this standpoint, the book joins other scholars in calling for urgent action in transforming the curriculum of business schools by adopting an interspecies perspective (e.g. Fennell et al., 2024; Tallberg et al., 2022).

These developments are not occurring in isolation. Across Europe and beyond, educators are experimenting with more-than-human approaches in management, tourism and sustainability. At institutions including Manchester Metropolitan University (UK), the Universities of Oulu, Lapland, Jyväskylä and Aalto (Finland) and Huron University College (Canada), this work ranges from individual sessions to dedicated modules and full programmes embedding animals and multispecies perspectives within business curricula. In Canada, recent programme developments at Huron have attracted national media attention, signalling growing public recognition of climate-responsive and ecologically oriented business education (CBC News, 2024). In Finland, multispecies perspectives on business and organising are also being advanced at the doctoral level through specialised seminars and research collaborations (Kataja, 2026). Taken together, these examples indicate an emerging pedagogical shift that this book seeks to support.

Such a transformation requires a genuinely multidisciplinary approach – one that draws on insights from diverse fields including critical animal studies, ecology, veterinary science, the social sciences and animal behaviour research. Integrating an interspecies perspective into the business curriculum invites a fundamental rethinking of what it means to teach business and, more importantly, what it means to teach responsibility, ethics and sustainability. Traditionally, these subjects have been grounded in a human-centric understanding of business and society, in which animals are framed primarily as symbols, inputs or resources – entities to be used, owned and discarded when no longer deemed useful (see Tallberg et al., 2022). An interspecies lens disrupts this anthropocentric model by encouraging educators and curriculum developers to view animals not as objects, but as participants in shaping organisational life, social values and our moral worlds. When animals are recognised as participants or stakeholders, we need to assume a moral responsibility to make their presence and interests visible within the organisational contexts, managerial practices and the theoretical concepts addressed in the business curriculum.

The pedagogical objective of transforming the curriculum extends beyond exposing harm and injustice. It also involves designing a curriculum that cultivates speculative thinking and fosters relational futures (see Gladkova & Matsuda, 2023). An interspecies curriculum must encourage students to imagine organisational systems where animals and other nonhuman species are all considered stakeholders. In doing so, the whole curriculum becomes a site for ethical experimentation and prefigurative practice. The goal is not merely to include animals in a single course, but to create opportunities to reflect on animals across diverse business fields, such as accounting, innovation, marketing and strategy. Rather than treating business systems and organising practices as fixed and immutable, this approach reframes them as collective, socio-ecological spaces embedded in a multispecies web of relations, enhancing learners' capacity to think, care and act beyond the human. By integrating animals into the business curriculum, we open the door to creative teaching and foster resilience in addressing deeply embedded assumptions about what learning is, who participates in it and who is excluded. Traditional business education emphasises human cognition, individual innovation and strategic control as the primary drivers of success. Yet lessons drawn from African forest beekeeping, animal work in tourism and wildlife conservation point to a radically different approach: One grounded in attentiveness, cooperation, relationships and responsiveness to changing local and global conditions.

In summary, teaching and learning is always an act of world-making. The concepts we foreground, the examples we choose and the absences we leave intact shape how future managers, policymakers and entrepreneurs come to understand their responsibilities (Cunliffe, 2004). If this book has shown anything, it is that animals have long been integral to business practice yet largely absent within business education (Colombino & Bruckner, 2023). Integrating them more deliberately into the curriculum is therefore not an optional enrichment, but part of rethinking what responsible management means in a multispecies world. In this sense, taking greater account of animals in our teaching and learning is not an add-on to responsible management, but a necessary step in building sustainable, ethical and multispecies futures.

References

CBC News. (2024, March 18). New Huron College degree aims to inspire students to tackle climate change. www.cbc.ca/news/canada/london/new-huron-college-degree-inspire-students-climate-change-1.7141038.

Colombino, A., & Bruckner, H. K. (2023). Hidden in plain sight: How (and why) to attend to the animal in human–animal relations. In A. Colombino & H. K. Bruckner (Eds), *Methods in human–animal studies* (pp. 1–29). Routledge. doi:10.4324/9781351018623-1.

Cunliffe, A. L. (2002). Reflexive dialogical practice in management learning. *Management Learning*, 33(1), 35–61. doi:10.1177/1350507602331002.

Cunliffe, A. L. (2004). On becoming a critically reflexive practitioner. *Journal of Management Education*, 28(4), 407–426. doi:10.1177/1052562904264440.

Fennell, D. A., Kline, C., Mkono, M., Grimwood, B. S. R., Sheppard, V. A., Dashper, K., … Madrid, R. (2024). Tourism, animals and the vacant niche: A scoping review and pedagogical agenda. *Current Issues in Tourism*, 27(22), 3820–3848. doi:10.1080/13683500.2023.2280704.

Fennell, D. A., & Sheppard, V. (2021). Tourism, animals and the scales of justice. *Journal of Sustainable Tourism*, 29(2–3), 314–335. doi:10.1080/09669582.2020.1768263.

García-Rosell, J.-C. (2012). Struggles over corporate social responsibility meanings in teaching practices: The case of hybrid problem-based learning. *Management Learning*, 44(5), 537–555. doi:10.1177/1350507612451228.

Gladkova, E., & Matsuda, N. (2023). Reimagining pigs: A multispecies, ecofeminist research method. *Australian Feminist Studies*, 38(115–116), 177–197. doi:10.1080/08164649.2024.2309619.

Höckert, E., Rantala, O., Garcia-Rosell Eskenazi, J., & Haanpää, M. (2019). Knowing with nature: The future of tourism education in the Anthropocene. *Journal of Teaching in Travel & Tourism*, 19(3), 264–265. doi:10.1080/15313220.2019.1620518.

Kataja. (2026). *Multispecies perspectives on business and organising (event listing).* https://kataja.eu/events/multispecies-perspectives-on-business-and-organizing-tba.

Kopnina, H. (2020). Of elephants and men. *Animal Sentience*, 5(28), 2. doi:10.51291/2377-7478.1553.

Kopnina, H., Roca, T., & Sorour, M. K. (2025). Integrating critical thinking about Sustainable Development Goals: Making biodiversity matter for business students. *Journal of Education for Sustainable Development*. doi:10.1177/09734082251336403.

Tallberg, L., Välikangas, L., & Hamilton, L. (2022). Animal activism in the business school: Using fierce compassion for teaching critical and positive perspectives. *Management Learning*, 53(1), 55–75. doi:10.1177/13505076211044612.

Tronto, J. (2020). *Moral boundaries: A political argument for an ethic of care*. Routledge. doi:10.4324/9781003070672.

Wadham, H., Hart, C., Hashmi, A., Kettleborough, H. M., Marron, R., Randles, S., … & Tucker, M. (2023). Sustainability education beyond the classroom: How the "exploding university" nurtures collective intelligence across local and global communities. In E. Meletiadou (Ed.), *Handbook of research on implications of sustainable development in higher education* (pp. 202–229). IGI Global.

14

CONCLUSION

Making animals conceptually central

This book began from the observation that animals are already central to business systems, but that they remain largely *absent* from how business and management are theorised, taught and governed (Pedersen, 2025). Across agriculture, tourism, conservation, digital media, healthcare, logistics and activism, animals are bred, worked, measured, marketed, regulated and killed within organisational arrangements structured primarily around human goals and needs. To ignore this is ethically troubling and throughout the preceding chapters we have argued that business has always been multi-species (Tallberg et al., 2024). The question now is not whether animals belong within business and management scholarship, but what changes when they are placed at the centre of business and management theory, education and practice.

By bringing animals into the centre of analysis, the book has shown how business practices actively organise animal lives through systems of breeding, care, surveillance, valuation, consumption, innovation and governance (Hamilton & McCabe, 2016). From elephants in conservation and tourism to broiler chickens in global supply chains, from horses in leisure industries to donkeys in rural economies, and from bees in forest systems to dogs in healthcare, animals appear not as abstract ethical problems but as participants in structured organisational worlds – whether within accounting systems, in company activism, or in other institutional arrangements.

The book has demonstrated that animals are already more than peripheral "stakeholders" and are constitutive of value creation, labour processes, market arrangements and institutional governance. Organising shapes their lives, but animal bodies, capacities, vulnerabilities and resistances also

DOI: 10.4324/9781003585541-20

shape organising. The book has thus rethought and reworked key foundational concepts in business and management. We have done so through a deliberate progression. Part I began by situating animals within foundational questions of sustainable development, welfare and religion. The book then moved to examine animals as social and economic actors within organisational systems. It then explored innovation, creativity and diversity through interspecies lenses, turned to questions of governance, value and activism and finally reflected on pedagogy as a site of institutional transformation. This movement – from moral foundations to economic participation, from creative possibility to institutional critique and ultimately to education – mirrors the argument that multispecies organising must be understood as systemic rather than peripheral. Across these movements, several core management concepts are reworked:

- *Work* is no longer exclusively human; it becomes interspecies labour shaped by cooperation, constraint and unequal power.
- *Consumption* becomes a multispecies metabolic process involving human bodies, animal bodies and planetary systems.
- *Innovation* shifts from human ingenuity alone to relational creativity that may draw upon, work with or exploit animal capacities.
- *Value* is exposed as a political and accounting construction that renders some forms of life legible and others invisible.
- *Diversity, equity and inclusion* expand beyond human categories to expose how species hierarchies intersect with race, class, gender and geography.
- *Governance and activism* become sites where species boundaries are reproduced and potentially transformed.

Throughout the book, animals function as an analytical lens. When we ask where animals are positioned within business systems, taken-for-granted distinctions between production and care, efficiency and ethics, means and ends begin to fracture and unravel. Anthropocentric assumptions, often operating silently, become visible, opening up space for a new ethical imagination. Importantly, the early chapters demonstrated that animals are not only economic actors but also moral and spiritual subjects embedded within systems of care, belief and obligation. Whether through religious reverence, welfare regimes or everyday practices of caregiving, animals are organised through ethical logics that exceed market calculation. Attending to these dimensions disrupts any simple framing of animals as units of production and reminds us that organising is always also a moral practice.

Pedagogy as institutional intervention

Another key thread running through the book is pedagogy. Business education shapes how future managers understand responsibility, value and harm.

If animals are absent from curricula, their lives become conceptually irrelevant to managerial thinking (Tallberg et al., 2022a). This book intervenes not only in scholarship but in teaching. By pairing familiar management concepts with interspecies cases and reflective questions, it offers a way of teaching that unsettles disciplinary comfort without abandoning conceptual rigour. It positions business education as a strategic site for cultivating new forms of ethical imagination and responsibility grounded in care (Kopnina, 2020). Rather than *adding* animals as another topic within sustainability or ethics modules, the book challenges the anthropocentric foundations of the curriculum itself.

In so doing, it treats pedagogy as a form of organisational and societal intervention. For educators, the book thus presents a challenge and an opportunity. If business education is world-making (Wadham et al., 2023), then teaching differently matters. The reflective questions embedded throughout are not merely exercises; they are invitations to notice absence, to interrogate metrics, to examine supply chains and to question what counts as legitimate business concern. Future pedagogical work is needed, but such work will not be easy in institutions shaped by growth-oriented logics. Yet it is precisely within such institutions that we reimagine the future.

Structural critique and economic reframing

The work outlined in the book is necessarily partial. Many species, sectors and geographical contexts remain underexplored. This book builds on and is indebted to a growing body of work in Animal Organisation Studies and related fields that has already unsettled anthropocentric assumptions within management and organisation theory (Tallberg & Hamilton, 2022). Our contribution has been to synthesise these strands, extend them across a wide range of core management concepts and bring them explicitly into conversation with pedagogy and institutional design. Building on this foundation, we argue that multispecies research must now engage more directly with the conceptual core of business and management scholarship rather than remaining at its margins.

Across the chapters, a consistent theme emerges: Responsibility cannot stop at the human. Organisational decisions distribute vulnerability across species (Blattner et al., 2020; Coulter, 2016). Animals are positioned within systems of dependency without representation or meaningful consent. These distributions are not accidental. They are produced by growth-driven economic systems, global supply chains and accounting practices that treat animal life primarily as a resource. They are also sustained by public policy decisions, regulatory frameworks and governance arrangements that normalise particular forms of interspecies dependency while rendering others unthinkable. Responsibility is therefore shared across corporations, governments and civil society actors, rather than residing solely within firms. Anthropocentrism is not separate from these arrangements. It is built into both economic and political systems. It shapes how markets operate and how policies are designed. Market priorities such as efficiency, profit and growth position animals as resources to be managed, while political decisions – including laws, subsidies and trade agreements – help maintain and legitimise these arrangements. In this way, market and political logics work together to stabilise and reproduce the status quo of interspecies exploitation. Rethinking responsibility, therefore, requires more than ethical reflection. It demands structural change in the economic and political arrangements that sustain interspecies exploitation. This raises serious questions for stakeholder models and growth-oriented orthodoxies that treat expansion as inherently good (Celermajer et al., 2023).

The collaborative structure of the book – developed through sustained dialogue with expert collaborators across diverse contexts – is itself part of its argument. Knowledge production here has been relational: Sometimes misaligned, occasionally uncomfortable and deliberately slow. If business organises animals through relations of care, control and dependency, then scholarship, too, is an organising practice. This book, therefore, models a multispecies approach to knowledge – one that experiments with co-creation, acknowledges partial perspectives and resists the individualised authorship norms of mainstream management research. By individualised authorship norms, we refer to the common practice of producing chapters or articles independently, with limited dialogue among contributors and clear ownership over separate intellectual territories. We deliberately worked against this model. Chapters were developed in conversation with one another and ideas evolved through shared reflection rather than isolated writing. We see individualised authorship as problematic because it can reproduce competitive academic cultures and limit the kinds of relational thinking that multispecies scholarship requires.

It also reflects the coexistence of different ontological and epistemological commitments among our collaborators. Although we share a concern for animals, contributors approach questions of agency, justice, reform and transformation from perspectives that are not always fully compatible. We see this diversity not as a problem to be solved, but as an inevitable and productive feature of a field that is still developing. In this sense, Animal Organisation Studies is not only expanding its object of analysis: It is also reconsidering how scholarship is practised. This same commitment also shaped our inclusion of both early-career and more established researchers. Supporting emerging scholars in this field is not simply a matter of inclusivity, though that matters. It is a recognition that reshaping business and management scholarship requires nurturing new voices and intellectual risk-taking. Modelling this intergenerational collaboration is part of the book's wider argument about how fields evolve and how responsibility is shared. Throughout the book, we use "we" intentionally to signal this shared authorship. The term refers to us as the authors of the volume, whose thinking has been shaped through engagement with each other and our expert collaborators rather than developed in isolation.

Productive tensions in multispecies organising

Once animals are recognised as participants in business systems, familiar arrangements no longer appear inevitable (Coulter, 2016). Industrial farming, animal-based tourism, laboratory testing, conservation governance, social media representation and corporate activism can all be reimagined. Some chapters gesture towards reform within existing systems; others suggest more transformative change. The book does not propose a single pathway. Rather, it equips readers with conceptual tools for thinking differently about organising in multispecies worlds. These differences are not weaknesses but productive tensions. The book does not resolve debates between reform and abolition, between welfare and liberation, or between working within institutions and challenging them from outside. Instead, it makes these tensions visible as part of the ongoing political and ethical work of multispecies organising. It therefore foregrounds a set of analytical questions that can guide future research and practice:

- Who bears vulnerability within this system and how is that vulnerability distributed across species?
- How is animal life valued, measured, priced or erased within organisational practices?
- What forms of interspecies labour are rendered invisible?
- Where are dependency, asymmetry and non-consensual relations structured into business models?
- What alternative forms of organising might reduce harm or redistribute responsibility?

These are not abstract ethical exercises (Tallberg et al., 2022b). They are key interventions that destabilise narrow stakeholder models and growth-oriented logics. By foregrounding animals, the book reframes strategy, governance, innovation and accountability as multispecies matters rather than exclusively human concerns.

We have been explicit about the book's limits. It is shaped by particular collaborations, by uneven research visibility and by structural constraints within academia. It cannot represent all species, regions or ethical positions. Yet conceptual marginalisation should not be mistaken for material insignificance. Interspecies organising has never been peripheral to economic life. Instead, dominant theoretical frameworks have relegated it to the margins.

As this book has shown, animals are deeply embedded in global supply chains, planetary crises and everyday economic practices. Climate change, biodiversity loss, zoonotic disease and food system instability are inseparable from how business systems organise animal life. To engage animals seriously is not a diversion from pressing business challenges: It is a way of confronting them directly.

If animals are constitutive participants in business systems, then organising must be approached differently. This concerns not only scholarship and pedagogy, but also those who design, govern and manage organisations. Decisions about sourcing, innovation, accounting and growth do more than shape operations; they organise interspecies vulnerability and distribute responsibility across species.

This recognition does not produce a single managerial formula. Instead, it shifts orientation: From treating animals as externalities to acknowledging them as participants and stakeholders in organisational systems (Sheppard & Fennell, 2019; Smart, 2022; Tallberg et al., 2022b). It may lead to reform within existing models or to more far-reaching transformation. The key point is that organising operates as a multispecies practice shaped by broader political and institutional arrangements.

An invitation to reconstitute business and management theory

The book opened by noting that animals already occupy a central place in business systems, yet remain largely invisible within business and management theory. It ends by insisting that this *absence* is not incidental but structural (Tallberg et al., 2024). The conceptual architecture of business has been built on anthropocentric exclusions that render animal lives marginal, instrumental or invisible. To make animals visible is not to widen the field slightly. Rather, it is to expose the limits of its foundations. Confronting the structural exclusions on which management theory has been built means it cannot simply extend existing categories. It must interrogate how its core assumptions about work, value, innovation, governance and responsibility have been constructed through species hierarchies.

The shift from inclusion to reconstitution reframes not only business organisations, but the broader institutional architectures that structure human–animal relations, including regulatory systems, financial regimes and public policy. For Animal Organisation Studies, this means moving beyond isolated case studies towards sustained engagement with core theories of strategy, finance, operations and organisational design – asking how these frameworks must shift when animals are understood not as externalities, but as participants within organisational systems. It means developing methodologies capable of tracing multispecies power relations across global supply chains and financial architectures. It means experimenting pedagogically within institutions that may resist such change, while remaining attentive to the risks of co-option, where the language of inclusion is absorbed without structural transformation. The work ahead is therefore intellectual, institutional and political; not an extension of existing business and management theory, but an ongoing reconsideration of its foundations.

References

Blattner, C. E., Coulter, K., & Kymlicka, W. (2020). *Animal labour: A new frontier of interspecies justice?* Oxford University Press.

Celermajer, D., Schlosberg, D., Rickards, L., Stewart-Harawira, M., Thaler, M., Tschakert, P., & Verlie, B. (2023). *Multispecies justice*. Cambridge University Press.

Coulter, K. (2016). *Animals, work and the promise of interspecies solidarity*. Palgrave Macmillan.

Hamilton, L., & McCabe, D. (2016). "It's just a job:" Understanding emotion work, animal suffering and resistance. *Organization*, 23(3), 330–349. doi:10.1177/1350508414559987.

Kopnina, H. (2020). Of elephants and men. *Animal Sentience*, 5(28), 2. doi:10.51291/2377-7478.1553.

Pedersen, H. (2025). Post-anthropocentric pedagogies: Purposes, practices and insights for higher education. *Teaching in Higher Education*, 30(2), 344–358. doi:10.1080/13562517.2023.2222087.

Sheppard, V. A., & Fennell, D. A. (2019). Progress in tourism public sector policy: Toward an ethic for non-human animals. *Tourism Management*, 73, 134–142. doi:10.1016/j.tourman.2018.11.017.

Smart, J. (2022). Animals as stakeholders. In N. Thomas (Ed.), *Animals and business ethics*. The Palgrave Macmillan animal ethics series (pp. 297–324). Palgrave Macmillan. doi:10.1007/978-3-030-97142-7_14.

Tallberg, L., & Hamilton, L. (2022). *The Oxford handbook of animal organization studies*. Oxford University Press.

Tallberg, L., Välikangas, L., & Hamilton, L. (2022a). Animal activism in the business school: Using fierce compassion for teaching critical and positive perspectives. *Management Learning*, 53(1), 55–75. doi:10.1177/13505076211044612.

Tallberg, L., García-Rosell, J. C., & Haanpää, M. (2022b). Human–animal relations in business and society: Advancing the feminist interpretation of stakeholder theory. *Journal of Business Ethics*, 180(1), 1–16. doi:10.1007/s10551-021-04820-2.

Tallberg, L., Huopalainen, A., & García-Rosell, J. C. (2024). Beyond anthropocentrism: A call to action for multispecies inclusivity. *Business and Society*. doi:10.1177/00076503241271254.

Wadham, H., Hart, C., Hashmi, A., Kettleborough, H. M., Marron, R., Randles, S., … & Tucker, M. (2023). Sustainability education beyond the classroom: How the "exploding university" nurtures collective intelligence across local and global communities. In E. Meletiadou (Ed.), *Handbook of research on implications of sustainable development in higher education* (pp. 202–229). IGI Global.

INDEX

For Product Safety Concerns and Information please contact our EU
representative GPSR@taylorandfrancis.com
Taylor & Francis Verlag GmbH, Kaufingerstraße 24, 80331 München, Germany

www.ingramcontent.com/pod-product-compliance
Ingram Content Group UK Ltd.
Pitfield, Milton Keynes, MK11 3LW, UK
UKHW020150200926
475702UK00016B/2644
* 9 7 8 1 0 3 2 9 5 5 6 5 0 *